Fig. 1

Fig. 4

Fig. 3

Fig. 2.

Engd L. S. Punderson

1. Telescopic view of the full Moon.
2. do do of a part of the Moon near quadrature.
3. Telescopic view of Saturn & his rings.
4. do of Jupiter & his Moons.

AN INTRODUCTION TO ASTRONOMY:

DESIGNED AS A

TEXT-BOOK

FOR THE USE OF

STUDENTS IN COLLEGE.

BY

DENISON OLMSTED, LL. D.,

LATE PROFESSOR OF NATURAL PHILOSOPHY AND ASTRONOMY IN YALE COLLEGE.

REVISED

BY E. S. SNELL, LL.D.,

PROFESSOR OF MATHEMATICS AND NATURAL PHILOSOPHY IN AMHERST COLLEGE.

NEW YORK:

COLLINS & BROTHER,

NO. 84 LEONARD STREET.

1865.

Entered according to Act of Congress, in the year 1844,
By DENISON OLMSTED,
In the Clerk's Office of the District Court of Connecticut.

Revised Edition.
Entered according to Act of Congress, in the year 1861,
By JULIA M. OLMSTED,
For the Children of Denison Olmsted, deceased,
In the Clerk's Office of the District Court of the District of Connecticut.

Rennie, Shea & Lindsay,
stereotypers and electrotypers,
81, 83, & 85 CENTRE-STREET,
New York.

C. A. ALVORD,
Printer,
15 Vandewater-street, New York.

PREFACE TO THE REVISED EDITION.

This work was revised by its author in 1854, at which time were introduced notices of the recent discoveries in Astronomy. The changes made in the present edition are mostly of a different character. While there have been added brief accounts of still later discoveries, and of new methods of observation, it has been especially my aim to give more clearness to certain descriptions and demonstrations, which my experience as a teacher has shown me to be perplexing to the learner. The discussion of central forces is entirely remodeled, and the great law of gravitation throughout the solar system is deduced, by a more direct and less cumbrous mode of reasoning, from the three laws of Kepler, which are taken as facts established by observation.

To those numerous teachers who approve the general design of Professor Olmsted in the preparation of this work, as set forth in his preface, and who have themselves tested the utility of that design, it is hoped that the changes now made will commend themselves as improvements.

E. S. SNELL.

Amherst College, March, 1861.

AUTHOR'S PREFACE.

Nearly all who have written Treatises on Astronomy, designed for young learners, appear to have erred in one of two ways; they have either disregarded demonstrative evidence, and relied on mere popular illustration, or they have exhibited the elements of the science in naked mathematical formulæ. The former are usually diffuse and superficial; the latter, technical and abstruse.

In the following Treatise, we have endeavored to unite the advantages of both methods. We have sought, first, to establish the great principles of astronomy on a mathematical basis; and, secondly, to render the study interesting and intelligible to the learner, by easy and familiar illustrations. We would not encourage any one to believe that he can enjoy a full view of the grand edifice of astronomy, while its noble foundations are hidden from his sight; nor would we assure him that he can contemplate the structure in its true magnificence, while its basement alone is within his field of vision. We would, therefore, that the student of astronomy should confine his attention neither to the exterior of the building, nor to the mere analytic investigation of its structure. We would desire that he should not only study it in models and diagrams, and mathematical formulæ, but should at the same time acquire a love of nature herself, and cultivate the habit of raising his views to the grand originals. Nor is the effort to form a clear conception of the motions and dimensions of the heavenly bodies, less favorable to the improvement of the intellectual powers, than the study of pure geometry.

But it is evidently possible to follow out all the intricacies of an analytical process, and to arrive at a full conviction of the great truths of astronomy, and yet know very little of nature. According to our experience, however, but few students in the course of a liberal education will feel satisfied with this. They do not need so much to be convinced that the assertions of astronomers are true, as they desire to know what the truths are, and how they were ascertained; and they will derive from the study of astronomy little of that moral and intellectual elevation which they had anticipated, unless they learn to look upon the heavens with new views, and a clear comprehension of their wonderful mechanism.

Much of the difficulty that usually attends the early progress of the astronomical student, arises from his being too soon introduced to the most perplexing part of the whole subject,—the planetary motions. In this work, the consideration of these is for the most part postponed until the learner has become familiar with the artificial circles of the sphere, and conversant with the celestial bodies. We then first take the most simple view possible of the planetary motions by contemplating them as they really are in nature, and afterwards proceed to the more difficult inquiry, why they appear as they do. Probably no science derives such signal advantage from a happy arrangement, as astronomy;—an order, which brings out every fact or doctrine of the science just in the place where the mind of the learner is prepared to receive it.

ANALYSIS.

PRELIMINARY OBSERVATIONS.

PART I. THE EARTH.

Chapter I.—OF THE FIGURE AND SIZE OF THE EARTH, AND THE DOCTRINE OF THE SPHERE.

Chapter II.—DIURNAL REVOLUTION—ARTIFICIAL GLOBES—ASTRONOMICAL PROBLEMS.

Chapter IV.—Precession of the Equinoxes—Nutation—Aberration—Mean and True Places of the Sun.

Chapter V.—The Moon—Lunar Geography—Phases of the Moon—Her Revolutions.

INTRODUCTION TO ASTRONOMY.

PRELIMINARY OBSERVATIONS.

1. ASTRONOMY *is that science which treats of the heavenly bodies.*

More particularly, its object is to teach what is known respecting the Sun, Moon, Planets, Comets, and Fixed Stars; and also to explain the methods by which this knowledge is acquired. Astronomy is sometimes divided into Descriptive, Physical, and Practical. Descriptive Astronomy respects *facts;* Physical Astronomy, *causes;* Practical Astronomy, the *means of investigating the facts*, whether by instruments, or by calculation. It is the province of Descriptive Astronomy to observe, classify, and record all the phenomena of the heavenly bodies, whether pertaining to those bodies individually, or resulting from their motions and mutual relations. It is the part of Physical Astronomy to explain the causes of these phenomena, by investigating and applying the general laws on which they depend; especially by tracing out all the consequences of the law of universal gravitation. Practical Astronomy lends its aid to both the other departments.

2. Astronomy is the most ancient of all the sciences. At a period of very high antiquity, it was cultivated in Egypt, in Chaldea, in China, and in India. Such knowledge of the heavenly bodies as could be acquired by close and long-continued observation, without the aid of instruments, was diligently amassed; and tables of the celestial motions were constructed, which could be used in predicting eclipses, and other astronomical phenomena.

About 500 years before the Christian era, *Pythagoras*, of Greece, taught astronomy at the celebrated school at Crotona, and exhibited more correct views of the nature of the celestial motions than were entertained by any other astronomer of the ancient world. His views, however, were not generally adopted, but lay neglected for nearly 2000 years, when they were revived and established by Copernicus and Galileo. The most celebrated astronomical school of antiquity was at Alexandria, in Egypt, which was established and sustained by the Ptolemies (Egyptian princes), about 300 years before the Christian era. The employment of instruments for measuring angles, and the introduction of trigonometrical calculations to aid the naked powers of observation, gave to the Alexandrian astronomers great advantages over all their predecessors. The most able astronomer of the Alexandrian school was *Hipparchus*, who was distinguished above all the ancients for the accuracy of his astronomical measurements and determinations. The knowledge of astronomy possessed by the Alexandrian school, and recorded in the *Almagest*, or great work of *Ptolemy*, constituted the chief of what was known of our science during the middle ages, until the fifteenth and sixteenth centuries, when the labors of *Copernicus* of Prussia, *Tycho Brahe* of Denmark, *Kepler* of Germany, and *Galileo* of Italy, laid the solid foundations of modern astronomy. Copernicus expounded the true theory of the celestial motions; Tycho Brahe carried the use of instruments and the art of astronomical observation to a far higher degree of accuracy than had ever been done before; Kepler discovered the great laws of the planetary motions; and Galileo, having first enjoyed the aid of the telescope, made innumerable discoveries in the solar system. Near the beginning of the eighteenth century, *Sir Isaac Newton* discovered, in the law of universal gravitation, the great principle that governs the celestial motions; and recently, *La Place* has more fully completed what Newton began, having followed out all the consequences of the law of universal gravitation, in his great work, the Mécanique Céleste.

3. Among the ancients, astronomy was studied chiefly as subsidiary to astrology. ASTROLOGY *was the art of divining future events by the stars.* It was of two kinds, natural and judicial. *Natural Astrology* aimed at predicting remarkable occurrences in the natural world, as earthquakes, volcanoes, tempests, and pestilential diseases. *Judicial Astrology* aimed at foretelling the fates of individuals or of empires.

4. Astronomers of every age have been distinguished for their persevering industry, and their great love of accuracy. They have uniformly aspired to an exactness in their inquiries far beyond what is aimed at in most geographical investigations, satisfied with nothing short of numerical accuracy, wherever this is attainable; and years of toilsome observation, or laborious calculation, have been spent with the hope of attaining a few seconds nearer to the truth. Moreover, a severe but delightful labor is imposed on all who would arrive at a clear and satisfactory knowledge of the subject of astronomy. Diagrams, artificial globes, orreries, and familiar comparisons and illustrations, proposed by the author or the instructor, may afford essential aid to the learner, but nothing can convey to him a perfect comprehension of the celestial motions, without much diligent study and reflection.

5. In expounding the doctrines of astronomy, we do not, as in geometry, claim that every thing shall be proved as soon as asserted. We may first put the learner in possession of the leading facts of the science, and afterwards explain to him the methods by which those facts were discovered, and by which they may be verified; we may assume the principles of the true system of the world, and employ those principles in the explanation of many subordinate phenomena, while we reserve the discussion of the merits of the system itself, until the learner is extensively acquainted with astronomical facts, and therefore better able to appreciate the evidence by which the system is established.

6. The *Copernican System* is that which is held to be the true system of the world. It maintains (1), That the *apparent* diurnal revolution of the heavenly bodies, from east to west, is owing to the *real* revolution of the earth on its own axis from west to east, in the same time; and (2), That the sun is the center around which the earth and planets all revolve from west to east, contrary to the opinion that the earth is the center of motion of the sun and planets.

7. We shall treat, first, of the Earth in its astronomical relations; secondly, of the Solar System; and, thirdly, of the Fixed Stars.

PART I.—OF THE EARTH.

CHAPTER I.

OF THE FIGURE AND DIMENSIONS OF THE EARTH, AND THE DOCTRINE OF THE SPHERE.

8. *The figure of the earth is nearly globular.* This fact is known, first, by the circular form of its shadow cast upon the moon in a lunar eclipse; secondly, from analogy, each of the other planets being seen to be spherical; thirdly, by our seeing the tops of distant objects while the other parts are invisible, as the topmast of a ship, while either leaving or approaching the shore, or the lantern of a light-house, which, when first descried at a distance at sea, appears to glimmer upon the very surface of the water; fourthly, by the depression or *dip of the horizon* when the spectator is on an eminence; and, finally, by actual observations and measurements, made for the express purpose of ascertaining the figure of the earth, by means of which astronomers are enabled to compute the distances from the center of the earth of various places on its surface, which distances are found to be nearly equal.

Fig. 1.

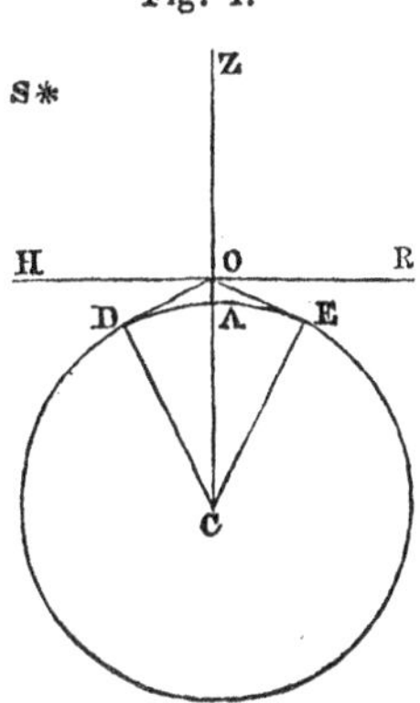

9. The *Dip of the Horizon*, is the apparent angular depression of the horizon, to a spectator elevated above the general level of the earth. The eye thus situated takes in more than a celestial hemisphere, the excess being the measure of the dip.

Thus, in Fig. 1, let AO represent

the height of a mountain, ZO the direction of the plumb-line, HOR a line passing through the station O, and at right angles to the plumb-line, C the center of the earth, DAE the portion of the earth's surface seen from O; OD, OE, lines drawn from the place of the spectator to the most distant parts of the horizon, and CD a radius of the earth. The dip of the horizon is the angle HOD or ROE. Now the angle made between the direction of the plumb-line and that of the extreme line of the horizon or the surface of the sea, namely, the angle ZOD, can be easily measured; and subtracting the right angle ZOH from ZOD, the remainder is the dip of the horizon, from which the length of the line OD may be calculated (see Art. 10), the height of the spectator, that is, the line OA, being known. This length, to whatever point of the horizon the line is drawn, is always found to be the same; and hence it is inferred, that the boundary which limits the view on all sides is a circle. Moreover, at whatever elevation the dip of the horizon is taken, in any part of the earth, the space seen by the spectator is always circular. Hence the surface of the earth is spherical.

10. The earth being a sphere, the dip of the horizon HOD = OCD. Therefore, to find the dip of the horizon corresponding to any given height AO* (the diameter of the earth being known), we have in the triangle OCD, the right angle at D, and the two sides CD, CO, to find the angle OCD. Therefore,

CO : rad. : : CD : cos. OCD.

Learning the dip corresponding to different altitudes, by giving to the line AO different values, we may arrange the results in a table.

---

* The learner will remark that the line AO, as drawn in the figure, is much larger in proportion to CA than is actually the case, and that the angle HOD is much too great for the reality. Such disproportions are very frequent in astronomical diagrams, especially when some of the parts are exceedingly small compared with others; and hence the diagrams employed in astronomy are not to be regarded as true *pictures* of the magnitudes concerned, but merely as representing their abstract geometrical *relations*.

Table showing the Dip of the Horizon at different elevations, from 1 *foot to* 100 *feet.**

Feet.	′ ″	Feet.	′ ″	Feet.	′ ″
1	0.59	13	3.33	26	5.01
2	1.24	14	3.41	28	5.13
3	1.42	15	3.49	30	5.23
4	1.58	16	3.56	35	5.49
5	2.12	17	4.03	40	6.14
6	2.25	18	4.11	45	6.36
7	2.36	19	4.17	50	6.58
8	2.47	20	4.24	60	7.37
9	2.57	21	4.31	70	8.14
10	3.07	22	4.37	80	8.48
11	3.16	23	4.43	90	9.20
12	3.25	24	4.49	100	9.51

Such a table is of use in estimating the altitude of a body above the horizon, when the instrument (as usually happens) is more or less elevated above the general level of the earth. For if it is a star whose altitude above the horizon is required, the instrument being situated at O (Fig. 1), the inquiry is, how far the star is elevated above the level HOR, but the angle taken is that above the visible horizon OD. The dip, therefore, or the angle HOD, corresponding to the height of the point O, must be subtracted, to obtain the true altitude. On the Peak of Teneriffe, a mountain 13,000 feet high, Humboldt observed the surface of the sea to be depressed on all sides nearly 2 degrees. The sun arose to him 12 minutes sooner than to an inhabitant of the plain; and from the plain, the top of the mountain appeared enlightened 12 minutes before the rising or after the setting of the sun.

11. The foregoing considerations show that the form of the earth is spherical; but more exact determinations prove that the earth, though nearly globular, is not exactly so: its diameter from the north to the south pole is about 26 miles less than through the equator, giving to the earth the form of an oblate spheroid,† or a flattened sphere resembling an orange. We

* This table includes the allowance for refraction.

† An *oblate spheroid* is the solid described by the revolution of an ellipse about its shorter axis.

shall reserve the explanations of the methods by which this fact is established, until the learner is better prepared than at present to understand them.

12. The mean or average *diameter of the earth* is 7912.4 miles, a measure which the learner should fix in his memory as a standard of comparison in astronomy, and of which he should endeavor to form the most adequate conception in his power. The circumference of the earth is about 25,000 miles (24857.5).* Although the surface of the earth is uneven, sometimes rising in high mountains, and sometimes descending in deep valleys, yet these elevations and depressions are so small in comparison with the immense volume of the globe, as hardly to occasion any sensible deviation from a surface uniformly curvilinear. The irregularities of the earth's surface in this view are no greater than the rough points on the rind of an orange, which do not perceptibly interrupt its continuity; for the highest mountain on the globe is only about five miles above the general level; and the deepest mine hitherto opened is only about half a mile.† Now $\frac{5}{7912}=\frac{1}{1582}$, or about one-sixteen-hundredth part of the whole diameter, an inequality which, in an artificial globe of eighteen inches diameter, amounts to only the eighty-eighth part of an inch.

13. The diameter of the earth, *considered as a perfect sphere*, may be determined by means of observations on a mountain of known elevation, seen in the horizon from the sea. Let BD (Fig. 2), be a mountain of known height a, whose top is seen in the horizon by a spectator at A, b miles from it. Let x denote the radius of the earth. Then $x^2+b^2=(x+a)^2=x^2+2ax+a^2$.

Fig. 2.

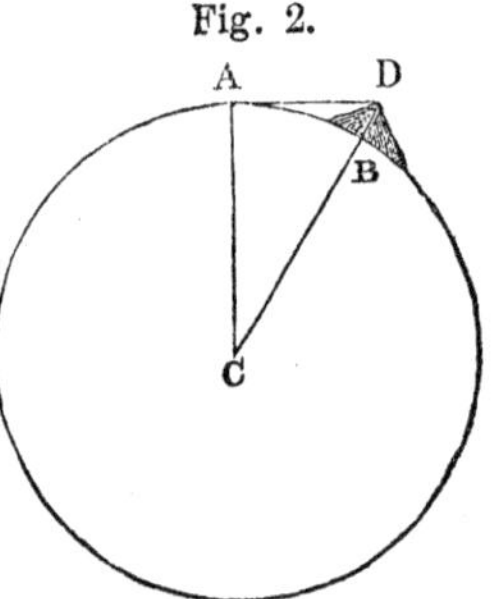

* It will generally be sufficient to treasure up in the memory the *round number*, but sometimes, in astronomical calculations, the more exact number may be required, and it is therefore inserted.

† Sir John Herschel.

Hence, $2ax = b^2 - a^2$, and $x = \frac{b^2 - a^2}{2a}$. For example, suppose the height of the mountain is just one mile; then it will be found, by observation, to be visible on the horizon at the distance of 89 miles $= b$. Hence, $\frac{b^2 - a^2}{2a} = \frac{(89)^2 - 1}{2} = \frac{7921 - 1}{2} = 3960 =$ radius of the earth, and $7920 =$ the earth's diameter.

14. Another method, and the most ancient, is to ascertain the distance on the surface of the earth, corresponding to a degree of latitude. Let us select two convenient places, one lying directly north of the other, whose difference of latitude is known. Suppose this difference to be 1° 30′, and the distance between the two places, as measured by a chain, to be 104 miles. Then, since there are 360 degrees of latitude in the entire circumference, 1° 30′ : 104 :: 360° : 24960. And $\frac{24960}{3.1416} = 7944$.

The foregoing *approximations* are sufficient to show that the earth is about 8,000 miles in diameter.

15. The greatest difficulty in the way of acquiring correct views in astronomy, arises from the erroneous notions that preoccupy the mind. To divest himself of these, the learner should conceive of the earth as a huge globe occupying a small portion

Fig. 3.

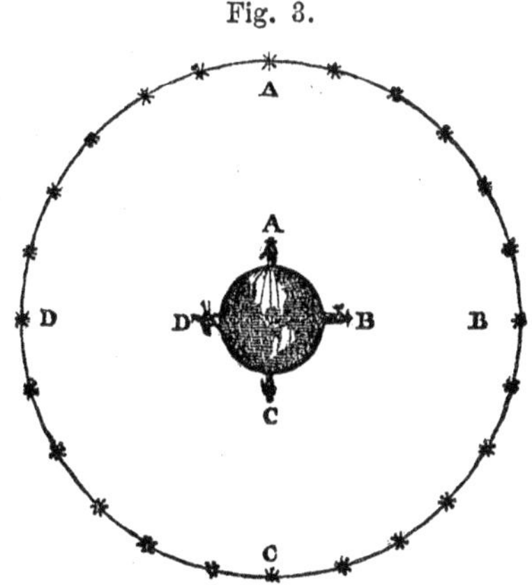

of space, and encircled on all sides with the starry sphere. He should free his mind from its habitual proneness to consider one part of space as naturally *up* and another *down*, and view himself as subject to a force which binds him to the earth as truly as though he were fastened to it by some invisible cords or wires, as the needle attaches itself to all sides of a spherical loadstone. He should dwell on this point until it appears to him as truly *up* in the direction of BB, CC, DD (Fig. 3), when he is at B, C, and D, respectively, as in the direction of AA when he is at A.

DOCTRINE OF THE SPHERE.

16. The definitions of the different lines, points, and circles, which are used in astronomy, and the propositions founded upon them, compose the *Doctrine of the Sphere.**

17. A section of a sphere by a plane cutting it in any manner, is a circle. *Great circles* are those which pass through the center of the sphere, and divide it into two equal hemispheres: *Small circles* are such as do not pass through the center, but divide the sphere into two unequal parts. The circumference of every circle, whether great or small, is divided into 360 equal parts called *degrees.* Hence, a degree is not a particular length of arc, but only a certain part of any whole circumference.

18. The *Axis* of a circle is a straight line passing through its center at right angles to its plane.

19. The *Pole* of a great circle is the point on the sphere where its axis cuts through the sphere. Every great circle has two poles, each of which is everywhere 90° from the great circle. For, the measure of an angle at the center of a sphere is the arc of a great circle intercepted between the two lines that contain the angle; and, since the angle made by the axis and any radius of the circle is a right angle, consequently its measure on

---

* It is presumed that many of those who read this work will have studied Spherical Geometry; but it is so important to the student of astronomy to have a clear idea of the circles of the sphere, that it is thought best to introduce them here.

the sphere, namely, the distance from the pole to the circumference of the circle, must be 90°. If two great circles cut each other at right angles, the poles of each circle lie in the circumference of the other circle. For each circle passes through the axis of the other.

20. All great circles of the sphere cut each other in two points diametrically opposite, and consequently their points of section are 180° apart. For the line of common section is a diameter in both circles, and therefore bisects both.

21. A great circle which passes through the pole of another great circle, cuts the latter at right angles. For, since it passes through the pole and the center of the circle, it must pass through the axis; which being at right angles to the plane of the circle, every plane which passes through it is at right angles to the same plane.

The great circle which passes through the pole of another great circle, and is at right angles to it, is called a *secondary* to that circle.

22. The angle made by two great circles on the surface of the sphere, is measured by the arc of another great circle, of which the angular point is the pole, being the arc of that great circle intercepted between those two circles. For this arc is the measure of the angle formed at the center of the sphere by two radii, drawn at right angles to the line of common section of the two circles, one in one plane and the other in the other, which angle is therefore that of the inclination of those planes.

23. In order to fix the position of any plane, either on the surface of the earth or in the heavens, both the earth and the heavens are conceived to be divided into separate portions by circles, which are imagined to cut through them in various ways. The earth, thus intersected, is called the *terrestrial*, and the heavens, the *celestial* sphere. The learner will remark that these circles have no existence in nature, but are mere landmarks, artificially contrived, for convenience of reference. On account of the immense distance of the heavenly bodies, they appear to us, wherever we are placed, to be fixed in the same concave surface, or celestial vault. The great circles of the globe, extended every way to meet the concave surface of the heavens, become circles of the celestial sphere.

24. The *Horizon* is the great circle which divides the earth into upper and lower hemispheres, and separates the visible heavens from the invisible. This is the *rational* horizon. The *sensible* horizon is a circle touching the earth at the place of the spectator, and is bounded by the line in which the earth and skies seem to meet. The sensible horizon is parallel to the rational, but is distant from it by the semi-diameter of the earth, or nearly 4,000 miles. Still, so vast is the distance of the starry sphere, that both these planes appear to cut that sphere in the same line; so that we see the same hemisphere of stars that we should see if the upper half of the earth were removed, and we stood on the rational horizon.

25. The poles of the horizon are the zenith and nadir. The *Zenith* is the point directly over our head, and the *Nadir* that directly under our feet. The plumb line is in the axis of the horizon, and consequently directed towards its poles.

Every place on the surface of the earth has its own horizon; and the traveller has a new horizon at every step, always extending 90 degrees from his zenith in all directions.

26. *Vertical circles* are those which pass through the poles of the horizon, perpendicular to it.

The *Meridian* is that vertical circle which passes through the north and south points.

The *Prime Vertical* is that vertical circle which passes through the east and west points.

27. As in geometry we determine the position of any point by means of rectangular co-ordinates, or perpendiculars drawn from the point to planes at right angles to each other, so in astronomy we ascertain the place of a body, as a fixed star, by taking its angular distance from two great circles, one of which is perpendicular to the other. Thus the horizon and the meridian, or the horizon and the prime vertical, are co-ordinate circles used for such measurements.

The *Altitude* of a body is its elevation above the horizon measured on a vertical circle.

The *Azimuth* of a body is its distance measured on the horizon from the meridian to a vertical circle passing through the body.

The *Amplitude* of a body is its distance on the horizon from the prime vertical to a vertical circle passing through the body.

Azimuth is reckoned 90° from either the north or south point; and amplitude 90° from either the east or west point. Azimuth and amplitude are mutually complements of each other. When a point is *on* the horizon it is only necessary to count the number of degrees of the horizon between that point and the meridian, in order to find its azimuth; but if the point is *above* the horizon, then its azimuth is estimated by passing a vertical circle through it, and reckoning the azimuth from the point where this circle cuts the horizon.

The *Zenith Distance* of a body is measured on a vertical circle passing through that body. It is the complement of the altitude.

28. The *Axis of the Earth* is the diameter on which the earth is conceived to turn in its diurnal revolution. The same line continued until it meets the starry concave, constitutes the *axis of the celestial sphere.*

The *Poles of the Earth* are the extremities of the earth's axis: the *Poles of the Heavens*, the extremities of the celestial axis.

29. The *Equator* is a great circle cutting the axis of the earth at right angles. Hence the axis of the earth is the axis of the equator, and its poles are the poles of the equator. The intersection of the plane of the equator with the surface of the earth, constitutes the *terrestrial*, and with the concave sphere of the heavens, the *celestial* equator. The latter, by way of distinction, is sometimes denominated the *equinoctial.*

30. The secondaries to the equator, that is, the great circles passing through the poles of the equator, are called *Meridians*, because that secondary which passes through the zenith of any place is the meridian of that place, and is at right angles both to the equator and the horizon, passing as it does through the poles of both. (Art. 21.) These secondaries are also called

Hour Circles, because the arcs of the equator intercepted between them are used as measures of time.

31. The *Latitude* of a place on the earth is its distance from the equator north or south. The *Polar Distance*, or angular distance from the nearest pole, is the complement of the latitude.

32. The *Longitude* of a place is its distance from some standard meridian, either east or west, measured on the equator. The meridian usually taken as the standard, is that of the Observatory of Greenwich, near London. If a place is directly *on* the equator, we have only to inquire how many degrees of the equator there are between that place and the point where the meridian of Greenwich cuts the equator. If the place is north or south of the equator, then its longitude is the arc of the equator intercepted between the meridian which passes through the place, and the meridian of Greenwich.

33. The *Ecliptic* is a great circle in which the earth performs its annual revolution around the sun. It passes through the center of the earth and the center of the sun. It is found by observation that the earth does not lie with its axis at right angles to the plane of the ecliptic, but that it is turned about 23½ degrees out of a perpendicular direction, making an angle with the plane itself of 66½°. The equator, therefore, must be turned the same distance out of a coincidence with the ecliptic, the two circles making an angle with each other of 23½° (23° 27′ 40″). It is particularly important for the learner to form correct ideas of the ecliptic, and of its relations to the equator, since to these two circles a great number of astronomical measurements and phenomena are referred.

34. The *Equinoctial Points*, or *Equinoxes*,* are the intersections of the ecliptic and equator. The time when the sun crosses the equator in returning northward is called the *vernal*, and in going southward the *autumnal* equinox. The vernal equinox occurs about the 21st of March, and the autumnal the 22d of September.

* The term Equinoxes strictly denotes the times when the sun arrives at the equinoctial points, but it is also frequently used to denote those points themselves.

35. The *Solstitial Points* are the two points of the ecliptic most distant from the equator. The times when the sun comes to them are called *solstices*. The summer solstice occurs about the 22d of June, and the winter solstice about the 22d of December.

The ecliptic is divided into twelve equal parts of 30° each, called *signs*, which, beginning at the vernal equinox, succeed each other in the following order:

Northern.		Southern.	
1. Aries	♈	7. Libra	♎
2. Taurus	♉	8. Scorpio	♏
3. Gemini	♊	9. Sagittarius	♐
4. Cancer	♋	10. Capricornus	♑
5. Leo	♌	11. Aquarius	♒
6. Virgo	♍	12. Pisces	♓

The mode of reckoning on the ecliptic, is by signs, degrees, minutes, and seconds. The sign is denoted either by its name or its number. Thus 100° may be expressed either as the 10th degree of Cancer, or as $3^s\ 10°$.

36. Of the various meridians, two are distinguished by the name of *Colures*. The *Equinoctial Colure* is the meridian which passes through the equinoctial points. The *Solstitial Colure* is the meridian which passes through the solstitial points. As the solstitial points are 90° from the equinoctial points, so the solstitial colure is 90° from the equinoctial colure. It is also at right angles, or a secondary to both the ecliptic and equator. For, like every other meridian, it is of course perpendicular to the equator, passing through its poles. Moreover, the equinox, being a point both in the equator and in the ecliptic, is 90° from the solstice, from the pole of the equator, and from the pole of the ecliptic. Hence the solstitial colure, which passes through the solstice and the pole of the equator, passes also through the pole of the ecliptic, being the great circle of which the equinox itself is the pole. Consequently the solstitial colure is a secondary to both the equator and the ecliptic. (See Arts. 19, 20, 21.)

37. The position of a celestial body is referred to the equator by its right ascension and declination. (See Art. 27.) *Right*

Ascension is the angular distance from the vernal equinox, measured on the equator. If a star is situated *on* the equator, then its right ascension is the number of degrees of the equator between the star and the vernal equinox. But if the star is north or south of the equator, then its right ascension is the arc of the equator intercepted between the vernal equinox and that secondary to the equator which passes through the star. *Declination* is the distance of a body from the equator, measured on a secondary to the latter. Therefore, right ascension and declination correspond to terrestrial longitude and latitude; right ascension being reckoned from the equinoctial colure, in the same manner as longitude is reckoned from the meridian of Greenwich. On the other hand, celestial longitude and latitude are referred, not to the equator, but to the ecliptic. *Celestial Longitude*, is the distance of a body from the vernal equinox reckoned on the ecliptic. *Celestial Latitude*, is the distance from the ecliptic measured on a secondary to the latter. Or, more briefly, Longitude is distance *on* the ecliptic: Latitude, distance *from* the ecliptic. The *North Polar Distance* of a star, is the complement of its declination.

Fig. 4.

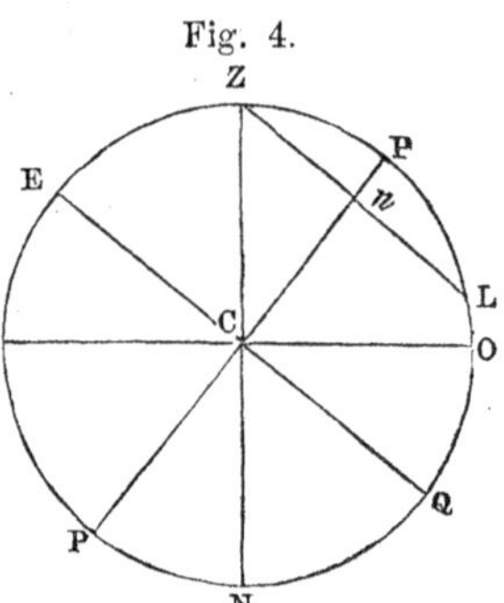

38. *Parallels of Latitude* are small circles parallel to the equator. They constantly diminish in size as we go from the equator to the pole, the radius being always equal to the cosine of the latitude. In fig. 4, let HO be the horizon, EQ the equator, PP the axis of the earth, ZN the prime vertical, and ZL a parallel of latitude of any place Z. Then ZE is the latitude (Art. 31), and ZP the complement of the latitude; but Z*n*, the radius of the parallel of latitude ZL, is the sine of ZP, and therefore the cosine of the latitude.

39. The *Tropics* are the parallels of latitude that pass through the solstices. The northern tropic is called the tropic of Cancer; the southern, the tropic of Capricorn.

40. The *Polar Circles* are the parallels of latitude that pass through the poles of the ecliptic, at the distance of 23½ degrees from the pole of the earth. (Art. 33.)

41. The earth is divided into five zones. That portion of the earth which lies between the tropics, is called the *Torrid Zone;* that between the tropics and the polar circles, the *Temperate Zones;* and that between the polar circles and the poles, the *Frigid Zones.*

42. The *Zodiac* is the part of the celestial sphere which lies about 8 degrees on each side of the ecliptic. This portion of the heavens is thus marked off by itself, because the planets are never seen further from the ecliptic than this limit.

43. *The elevation of the pole is equal to the latitude of the place.*

The arc PE (Fig. 4.)=ZO $\therefore$ PO=ZE, which equals the latitude.

44. *The elevation of the equator is equal to the complement of the latitude.*

ZH=90°. But ZE=Lat. $\therefore$ EH=90−Lat.=colatitude.

45. *The distance of any place from the pole (or the polar distance) equals the complement of the latitude.*

EP=90°. But EZ=Lat. $\therefore$ ZP=90−Lat.=colatitude.

CHAPTER II.

DIURNAL REVOLUTION—ARTIFICIAL GLOBES—ASTRONOMICAL PROBLEMS.

46. The apparent diurnal revolution of the heavenly bodies from east to west, is owing to the actual revolution of the earth on its own axis from west to east. If we conceive of a radius of the earth's equator extended until it meets the concave sphere of the heavens, then, as the earth revolves, the extremity of this line would trace out a curve on the face of the sky, namely, the celestial equator. In curves parallel to this, called the *circles of diurnal revolution*, the heavenly bodies actually *appear* to move, every star having its own peculiar circle. After the learner has first rendered familiar the real motions of the earth from west to east, he may then, without danger of misconception, adopt the common language, that all the heavenly bodies revolve around the earth once a day from east to west, in circles parallel to the equator and to each other.

47. The time occupied by a star in passing from any point in the meridian until it comes round to the same point again, is called a *sidereal day*, and measures the period of the earth's revolution on its axis. If we watch the returns of the same star from day to day, we shall find the intervals exactly equal to one another; that is, *the sidereal days are all equal.** Whatever star we select for the observation, the same result will be obtained. The stars, therefore, always keep the same relative position, and have a common movement round the earth,—a consequence that naturally flows from the hypothesis that their *apparent* motion is all produced by a single real motion, namely, that of the earth. The sun, moon, and planets, revolve in like manner, but their returns to the meridian are not, like those of the fixed stars, at exactly equal intervals.

48. The *appearances* of the diurnal motions of the heavenly bodies are different in different parts of the earth, since every place has its own horizon (Art. 15), and different horizons are variously inclined to each other. Let us suppose the spectator viewing the diurnal revolutions, successively, from several different positions on the earth.

49. If he is on the *equator*, his horizon passes through both poles; for the horizon cuts the celestial vault at 90 degrees in every direction from the zenith of the spectator; but the pole is likewise 90 degrees from his zenith, and consequently, the

---

* Allowance is here supposed to be made for the effects of precession, &c.

pole must be in his horizon. The celestial equator coincides with his prime vertical, being a great circle passing through the east and west points. Since all the diurnal circles are parallel to the equator, they are all, like the equator, perpendicular to his horizon. Such a view of the heavenly bodies is called a right sphere; or,

A Right Sphere *is one in which all the daily revolutions of the heavenly bodies are in circles perpendicular to the horizon.*

A right sphere is seen only at the equator. Any star situated in the celestial equator, would appear to rise directly in the east, to pass the meridian in the zenith of the spectator, and to set directly in the west. In proportion as stars are at a greater distance from the equator towards the pole, they describe smaller and smaller circles, until, near the pole, their motion is hardly perceptible. In a right sphere every star remains an equal time above and below the horizon; and since the times of their revolutions are equal, the velocities are as the lengths of the circles they describe. Consequently, as the stars are more remote from the equator towards the pole, their motions become slower, until, at the pole, if a star were there, it would appear stationary.

50. If the spectator advances one degree towards the north pole, his horizon reaches one degree beyond the pole of the earth, and cuts the starry sphere one degree below the pole of the heavens, or below the north star if that be taken as the place of the pole. As he moves onward towards the pole, his horizon continually reaches further and further beyond it, until, when he comes to the pole of the earth, and under the pole of the heavens, his horizon reaches on all sides to the equator, and coincides with it. Moreover, since all the circles of daily motion are parallel to the equator, they become, to the spectator at the pole, parallel to the horizon. This is what constitutes a parallel sphere. Or,

A Parallel Sphere *is that in which all the circles of daily motion are parallel to the horizon.*

51. To render this view of the heavens familiar, the learner should follow round in his mind a number of separate stars, one near the horizon, one a few degrees above it, and a third near the zenith. To one who stood upon the north pole, the stars of the northern hemisphere would all be perpetually in view when not obscured by clouds or lost in the sun's light, and none of those of the southern hemisphere would ever be seen. The sun would be constantly above the horizon for six months in the year, and the remaining six constantly out of sight. That is, at the pole, the days and nights are each six months long. The phenomena at the south pole are similar to those at the north.

52. A perfect parallel sphere can never be seen except at one of the poles,—a point which has never been actually reached by man; yet the British discovery-ships penetrated within a few degrees of the north pole, and of course enjoyed the view of a sphere nearly parallel.

53. As the circles of daily motion are parallel to the horizon of the pole, and perpendicular to that of the equator, so at all places between the two, the diurnal motions are oblique to the horizon. This aspect of the heavens constitutes an oblique sphere, which is thus defined:

An Oblique Sphere *is that in which the circles of daily motion are oblique to the horizon.*

Suppose for example the spectator is at the latitude of fifty degrees. His horizon reaches 50° beyond the pole of the earth, and gives the same apparent elevation to the pole of the heavens. It cuts the equator, and all the circles of daily motion, at an angle of 40°, being always equal to the co-altitude of the pole. Thus, let HO (Fig. 5), represent the horizon, EQ the equator, and PP′ the axis of the earth. Also, *ll*, *mm*, &c., parallels of latitude. Then the horizon of a spectator at Z, in latitude 50° reaches to 50° beyond the pole (Art. 50); and the angle ECH, is 40°. As we advance still further north, the elevation of the diurnal circles grows less

Fig. 5.

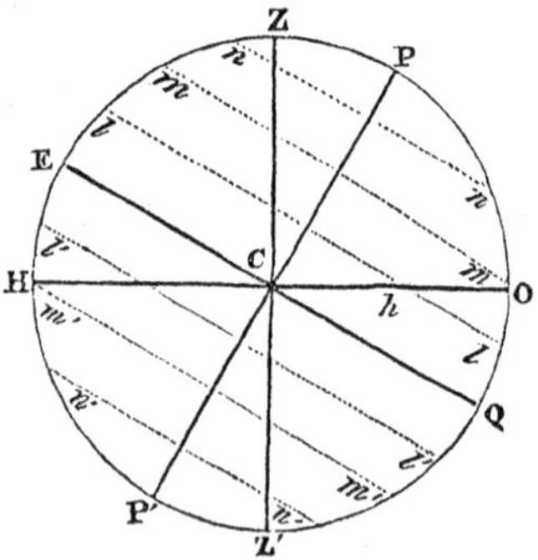

and less, and consequently the motions of the heavenly bodies more and more oblique, until finally, at the pole, where the latitude is 90°, the angle of elevation of the equator vanishes, and the horizon and the equator coincide with each other, as before stated.

54. *The* CIRCLE OF PERPETUAL APPARITION, *is the boundary of that space around the elevated pole, where the stars never set.* Its distance from the pole is equal to the latitude of the place. For, since the altitude of the pole is equal to the latitude, a star whose polar distance is just equal to the latitude, will, when at its lowest point, only just reach the horizon; and all the stars nearer the pole than this will evidently not descend so far as the horizon.

Thus, *mm* (Fig. 5), is the circle of perpetual apparition, between which and the north pole, the stars never set, and its distance from the pole OP is evidently equal to the elevation of the pole, and of course to the latitude.

55. In the opposite hemisphere, a similar part of the sphere adjacent to the depressed pole never rises. Hence,

The CIRCLE OF PERPETUAL OCCULTATION, *is the boundary of that space around the depressed pole, within which the stars never rise.* Thus, *m′m′* (Fig. 5), is the circle of perpetual occultation, between which and the south pole the stars never rise.

56. In an oblique sphere, the horizon cuts the circles of daily motion unequally. Towards the elevated pole, more than half the circle is above the horizon, and a greater and greater portion as the distance from the equator is increased, until finally, within the circle of perpetual apparition, the whole circle is above the horizon. Just the opposite takes place in the hemisphere next the depressed pole. Accordingly, when the sun is in the equator, as the equator and horizon, like all other great circles of the sphere, bisect each other, the days and nights are equal all over the globe. But when the sun is north of the equator, our days become longer than our nights, but shorter when the sun is south of the equator. Moreover, the higher the latitude, the greater is the inequality in the lengths of the days and nights. All these points will be readily understood by inspecting figure 5.

57. Most of the phenomena of the diurnal revolution can be explained, either on the supposition that the celestial sphere actually all turns around the earth once in 24 hours, or that this motion of the heavens is merely apparent, arising from the revolution of the earth on its axis in the opposite direction,—a motion of which we are insensible, as we sometimes lose the consciousness of our own motion in a ship or a steamboat, and observe all external objects to be receding from us with a common motion. Proofs entirely conclusive and satisfactory, establish the fact, that it is the earth and not the celestial sphere that turns; but these proofs are drawn from various sources, and the student is not prepared to appreciate their value, or even to understand some of them, until he has made considerable proficiency in the study of astronomy, and become familiar with a great variety of astronomical phenomena. To such a period of our course of instruction we therefore postpone the discussion of the hypothesis of the earth's rotation on its axis.

58. While we retain the same place on the earth, the diurnal revolution occasions no change in our horizon, but our horizon goes round as well as ourselves. Let us first take our station on the equator at sunrise; our horizon now passes through both the poles, and through the sun, which we are to conceive of as at a great distance from the earth, and therefore as cut, not by the terrestrial but by the celestial horizon. As the earth turns, the horizon dips more and more below the sun, at the rate of 15 degrees for every hour; and, as in the case of the polar star (Art. 50), the sun appears to rise at the same rate. In six hours, therefore, it is depressed 90 degrees below the sun, which brings us directly under the sun, which, for our present purpose, we may consider as having all the while maintained the same fixed position in space. The earth continues to turn, and in six hours more, it completely reverses the position of our horizon, so that the western part of the horizon which at sunrise was diametrically opposite to the sun now cuts the sun, and soon afterwards it rises above the level of the sun, and the sun sets. During the next twelve hours, the sun continues on the invisible side of the sphere, until the horizon returns to the position from which it started, and a new day begins.

59. Let us next contemplate the similar phenomena at the *poles*. Here the horizon, coinciding as it does with the equator, would cut the sun through its center, and the sun would appear to revolve along the surface of the sea, one half above and the other half below the horizon. This supposes the sun in its annual revolution to be at one of the equinoxes. When the sun is north of the equator, it revolves continually round in a path which, during a single revolution, appears parallel to the horizon, and it is constantly day; and when the sun is south of the equator, it is, for the same reason, continual night.

60. We have endeavored to conceive of the manner in which the apparent diurnal movements of the sun are *really* produced at two stations, namely, in the right sphere, and in the parallel sphere. These two cases being clearly understood, there will be little difficulty in applying a similar explanation to an oblique sphere.

ARTIFICIAL GLOBES.

61. Artificial globes are of two kinds, terrestrial and celestial. The first exhibits a miniature representation of the earth; the second, of the visible heavens; and both show the various circles by which the two spheres are respectively traversed. Since all globes are similar solid figures, a small globe, imagined to be situated at the center of the earth or of the celestial vault, may represent all the visible objects and artificial divisions of either sphere, and with great accuracy and just proportions, though on a scale greatly reduced. The study of artificial globes, therefore, cannot be too strongly recommended to the student of astronomy.*

* It were desirable, indeed, that every student of the science should have the celestial globe at least, constantly before him. One of a small size, as eight or nine inches, will answer the purpose, although globes of these dimensions cannot usually be relied on for nice measurements.

62. An artificial globe is encompassed from north to south by a strong brass ring, to represent the meridian of the place. This ring is made fast to the two poles, and thus supports the globe, while it is itself supported in a vertical position by means of a frame, the ring being usually let into a socket in which it may be easily slid, so as to give any required elevation to the pole. The brass meridian is graduated each way from the equator to the pole 90°, to measure degrees of latitude or declination, according as the distance from the equator refers to a point on the earth or in the heavens. The horizon is represented by a broad zone, made broad for the convenience of carrying on it a circle of azimuth, another of amplitude, and a wide space on which are delineated the signs of the ecliptic, and the sun's place for every day in the year; not because these points have any special connection with the horizon, but because this broad surface furnishes a convenient place for recording them.

63. *Hour Circles* are represented on the terrestrial globe by great circles drawn through the pole of the equator; but, on the celestial globe, corresponding circles pass through the poles of the ecliptic, constituting *circles of celestial latitude* (Art. 37), while the brass meridian, being a secondary to the equinoctial, becomes an hour circle of any star which, by turning the globe, is brought under it.

64. The *Hour Index* is a small circle described around the pole of the equator, on which are marked the hours of the day. As this circle turns along with the globe, it makes a complete revolution in the same time with the equator; or, for any less period, the same number of degrees of this circle and of the equator pass under the meridian. Hence the hour index measures arcs of right ascension. (Art. 37.)

65. The *Quadrant of Altitude* is a flexible strip of brass, graduated into ninety equal parts, corresponding in length to degrees on the globe, so that when applied to the globe and bent so as closely to fit its surface, it measures the angular distance between any two points. When the zero, or the point where the graduation begins, is laid on the pole of any great circle, the 90th degree will reach to the circumference of that circle, and being therefore a great circle passing through the pole of another great circle, it becomes a secondary to the latter. (Art. 21.) Thus the quadrant of altitude may be used as a secondary to any great circle on the sphere; but it is used chiefly as a secondary to the horizon, the point marked 90° being screwed fast to the pole of the horizon, that is, the zenith, and the other end, marked 0, being slid along between the surface of the sphere and the wooden horizon. It thus becomes a vertical circle, *on* which to measure the altitude of any star through which it passes, or *from* which to measure the azimuth of the star, which is the arc of the horizon intercepted between the meridian and the quadrant of altitude passing through the star. (Art. 27.)

66. To *rectify the globe for any place*, the north pole must be elevated to the latitude of the place (Art. 43); then the equator and all the diurnal circles will have their due inclination in respect to the horizon; and, on turning tho globo (tho celestial globe *west* and the terrestrial *east*), every point on either globe will revolve as the same point does in nature; and the relative situations of all places will be the same as on the respective native spheres.

PROBLEMS ON THE TERRESTRIAL GLOBE.

67. *To find the Latitude and Longitude of a place:* Turn the globe so as to bring the place to the brass meridian; then the degree and minute on the meridian directly over the place will indicate its latitude, and the point of the equator under the meridian, will show its longitude.

Ex. What are the Latitude and Longitude of the city of New York?

68. *To find a place having its latitude and longitude given:* Bring to the brass meridian the point of the equator corresponding to the longitude, and then at the degree of the meridian denoting the latitude, the place will be found.

Ex. What place on the globe is in Latitude 39 N. and Longitude 77 W.?

69. *To find the bearing and distance of two places:* Rectify the globe for one of the places (Art. 66); screw the quadrant of altitude to the zenith,* and let it pass through the other place. Then the azimuth will give the bearing of the second place from the first, and the number of degrees on the quadrant of altitude, multiplied by 69½ (the number of miles in a degree), will give the distance between the two places.

Ex. What is the bearing of New Orleans from New York, and what is the distance between these places?

70. *To determine the difference of time in different places:* Bring the place that lies eastward of the other to the meridian, and set the hour index at XII. Turn the globe eastward until the other place comes to the meridian, then the index will show the hour at the second place when it is noon at the first.

Ex. When it is noon at New York, what time is it at London?

71. *The hour being given at any place, to tell what hour it is in any other part of the world:* Find the difference of time between the two places (Art. 70), and, if the place whose time is required is eastward of the other, add this difference to the given time, but if westward, subtract it.

Ex. What time is it at Canton, in China, when it is 9 o'clock A. M. at New York?

72. *To find the antœci,† the periœci,‡ and the antipodes§ of any place:* Bring the given place to the meridian; then, under the meridian, in the opposite hemisphere, in the same degree of latitude, will be found the antœci. The same place remaining under the meridian, set the index to XII, and turn the globe until the other XII is under the index; then the periœci will be on the meridian, under the same degree of latitude with the given place, and the antipodes will be under the meridian, in the same latitude, in the opposite hemisphere.

Ex. Find the antœci, the periœci, and the antipodes of the citizens of New York.

The antœci have the same hour of the day, but different

---

* The zenith will of course be the point of the meridian over the place.

† αντι οικος. ‡ περι οικος. § αντι πȣς.

seasons of the year; the periœci have the same seasons, but opposite hours; and the antipodes have both opposite hours and opposite seasons.

73. *To rectify the globe for the sun's place:* On the wooden horizon, find the day of the month, and against it is given the sun's place in the ecliptic, expressed by signs and degrees.* Look for the same sign and degree on the ecliptic, bring that point to the meridian, and set the hour index to XII. To all places under the meridian it will then be noon.

Ex. Rectify the globe for the sun's place on the 1st of September.

74. *The latitude of the place being given, to find the time of the sun's rising and setting on any given day at that place:* Having rectified the globe for the latitude (Art. 66), bring the sun's place in the ecliptic to the graduated edge of the meridian, and set the hour index to XII; then turn the globe so as to bring the sun to the eastern and then to the western horizon, and the hour index will show the times of rising and setting respectively.

Ex. At what time will the sun rise and set at New Haven, Lat. 41° 18′, on the 10th of July?

PROBLEMS ON THE CELESTIAL GLOBE.

75. *To find the Declination and Right Ascension of a heavenly body:* Bring the place of the body (whether the sun or a star) to the meridian. Then the degree and minute standing over it will show its declination, and the point of the equinoctial under the meridian will give its right ascension. It will be remarked, that the declination and right ascension are found in the same manner as latitude and longitude on the terrestrial globe. Right ascension is expressed either in degrees or in hours; both being reckoned from the vernal equinox (Art. 37).

Ex. What is the declination and right ascension of the bright star Lyra?—also of the sun on the 5th of June?

---

* The larger globes have the day of the month marked against the corresponding sign on the ecliptic itself.

76. *To represent the appearance of the heavens at any time:* Rectify the globe for the latitude, bring the sun's place in the ecliptic to the meridian, and set the hour index to XII.; then turn the globe westward until the index points to the given hour, and the constellations would then have the same appearance to an eye situated at the center of the globe, as they have at that moment in the sky.

Ex. Required the aspect of the stars at New Haven, Lat. 41° 18′, at 10 o'clock, on the evening of December 5th.

77. *To find the altitude and azimuth of any star:* Rectify the globe for the latitude and the sun's place, and let the quadrant of altitude be screwed to the zenith, and be made to pass through the star. The arc on the quadrant, from the horizon to the star, will denote its altitude, and the arc of the horizon from the meridian to the quadrant, will be its azimuth.

Ex. What are the altitude and azimuth of Sirius (the brightest of the fixed stars) on the 25th of December at 10 o'clock in the evening, in Lat. 41°?

78. *To find the angular distance of two stars from each other:* Apply the zero mark of the quadrant of altitude to one of the stars, and the point of the quadrant which falls on the other star, will show the angular distance between the two.

Ex. What is the distance between the two largest stars of the Great Bear?*

79. *To find the sun's meridian altitude, the latitude and day of the month being given:* Having rectified the globe for the latitude (Art. 66), bring the sun's place in the ecliptic to the meridian, and count the number of degrees and minutes between that point of the meridian and the zenith. The complement of this arc will be the sun's meridian altitude.

Ex. What is the sun's meridian altitude at noon on the 1st of August, in Lat. 41° 18′?

* These two stars are sometimes called "the Pointers," from the line which passes through them being always nearly in the direction of the north star. The angular distance between them is about 5°, and may be learned as a standard for reference in estimating, by the eye, the distance between any two points on the celestial vault.

CHAPTER III.

OF PARALLAX, REFRACTION, AND TWILIGHT.

80. Parallax *is the apparent change of place which bodies undergo by being viewed from different points.* This change is called *diurnal* parallax, when one point is at the surface of the

Fig. 6.

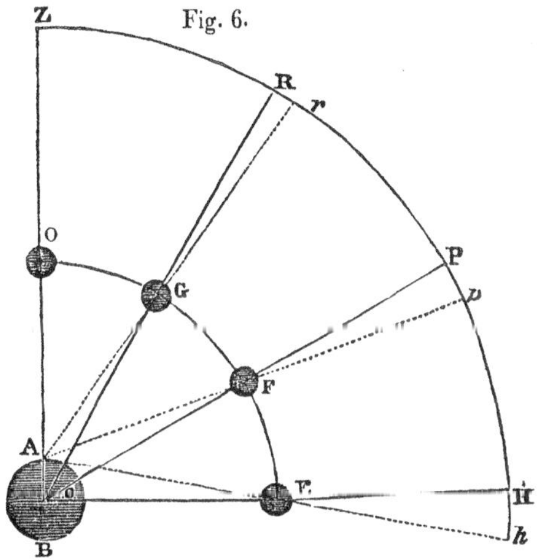

earth, and the other at the center. Thus (Fig. 6), an observer on the surface at A would see the moon F at p on the sky, while from the center C, it would appear at P; or, if the moon were at E, H would be its apparent place as seen from the center, and h from the surface. The angle AEC, or AFC, is the *diurnal parallax;* and Hh or Pp is the parallactic arc. It appears, therefore, that the diurnal parallax of a body is the angle at that body, subtended by the earth's radius.

81. Since then a heavenly body is liable to be referred to different points on the celestial vault, when seen from different parts of the earth, and thus some confusion occasioned in the determination of points on the celestial sphere, astronomers have agreed to consider the true place of a celestial object to be that where it would appear if seen from the center of the earth. The doctrine of parallax teaches how to reduce observations made at any place on the surface of the earth, to such as they would be if made from the center.

82. The angle AEC is called the horizontal parallax, which may be thus defined. *Horizontal parallax* is the change of position which a celestial body, appearing in the horizon as seen from the surface of the earth, would assume if viewed from the earth's center. It is the angle subtended by the earth's radius, when viewed perpendicularly from the body. If we consider any one of the triangles represented in the figure, ACG for example,

$$\text{Sin AGC} : \text{Sin GAZ} \ (= \text{Sin GAC}) :: \text{AC} : \text{CG};$$

$$\therefore \text{Sin Parallax} = \frac{\text{Sin GAZ} \times \text{AC}}{\text{CG}} \propto \frac{\text{Sin GAZ}}{\text{CG}}.$$

Hence the sine of the angle of parallax, or (since the angle of parallax is always very small)* *the parallax itself varies as the sine of the zenith distance of the body directly, and the distance of the body from the center of the earth inversely.* Parallax, therefore, increases as a body approaches the horizon (but increasing with the sines, it increases much slower than in the simple ratio of the distance from the zenith), and diminishes, as the distance from the spectator increases. Again, since the parallax AGC is as the sine of the zenith distance, let P represent the horizontal parallax, and P′ the parallax at any altitude; then,

$$P' : P :: \text{sin zenith dist.} : \text{sin } 90^\circ \therefore P = \frac{P'}{\text{sin zen. dist.}}.$$

Hence, the horizontal parallax of a body equals its parallax at any altitude, divided by the sine of its distance from the zenith.

83. From observations, therefore, on the parallax of a body at any elevation, we are enabled to find the angle subtended

---

* The moon, on account of its nearness to the earth, has the greatest horizontal parallax of any of the heavenly bodies; yet this is less than 1° (being 57′), while the greatest parallax of any of the planets does not exceed 30″. The difference in an arc of 1°, between the length of the arc and the sine, is only 0″.18.

by the semi-diameter of the earth as seen perpendicularly from the body. Or if the horizontal parallax is given, the parallax at any altitude may be found, for

$$P' = P \times \sin \text{zenith distance}.$$

Hence, in the zenith the parallax is nothing, and is at its maximum in the horizon.

84. It is evident from the figure, that the effect of parallax upon the place of a celestial body is *to depress it.* Thus, in consequence of parallax, E is depressed by the arc H*h*; F by the arc P*p*; G by the arc R*r*; while O sustains no change. Hence, in all calculations respecting the altitude of the sun, moon, or planets, the amount of parallax is to be added; the stars, as we shall see hereafter, have no sensible parallax. As the depression which arises from parallax is in the direction of a vertical circle, a body, when on the meridian, has only a parallax in declination; but in other situations, there is at the same time a parallax in declination and right ascension; for its direction being *oblique* to the equinoctial, it can be resolved into two parts, one of which (the declination) is perpendicular, and the other (the right ascension) is parallel to the equinoctial.

85. *The mode of determining the horizontal parallax*, is as follows:

Let O, O′ (Fig. 7), be two places on the earth, situated under the same meridian, at a great distance from each other; one place, for example, at the Cape of Good Hope, and the other in the north of Europe. The latitude of each place being known, the arc of the meridian OO′ is known, and the angle OCO′ also is known. Let the celestial body M (the moon for example), be observed simultaneously at O and O′, and its zenith distance at each place

Fig. 7.

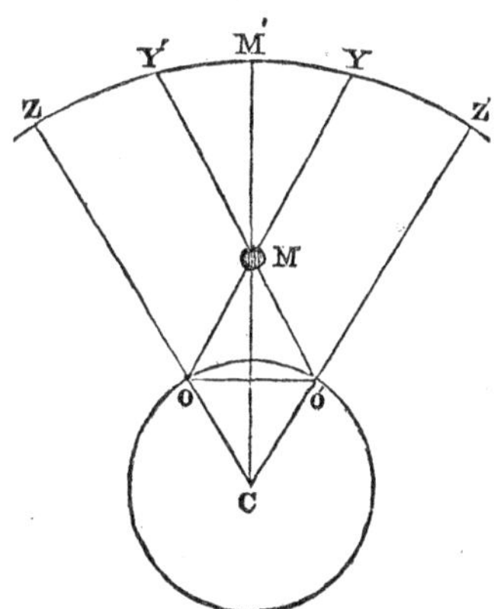

accurately taken, namely, ZY and Z′Y′; then the angles ZOM and Z′O′M, and of course their supplements COM, CO′M are found. Join OO′, and then in the isosceles triangle OCO′, determine the angles O, O′, and the side OO′; subtract O′OC from MOC, and OO′C from MO′C, and then in the triangle MOO′, we can calculate MO and MO′. Finally, in the triangle MOC, two known sides and their contained angle will furnish the angle OMC (which is the parallax at the station O), and the distance MC. To obtain the horizontal parallax P, we have (Art. 82),

$$P=\frac{OMC}{\text{Sin } ZOY}.$$

On this principle, the horizontal parallax of the moon was determined by La Caille and La Lande, two French astronomers, one stationed at the Cape of Good Hope, the other at Berlin; and in a similar way the parallax of Mars was ascertained, by observations made simultaneously at the Cape of Good Hope and at Stockholm.

86. On account of the great distance of the sun, his horizontal parallax, which is only 8″.6, cannot be accurately ascertained by this method. It can, however, be determined by means of the transits of Venus, a process to be described hereafter.

87. The determination of the horizontal parallax of a celestial body is an element of great importance, since it furnishes the means of estimating the distance of the body from the center of the earth. Thus, if the angle AEC (Fig. 6) be found, the radius of the earth AC being known, we have in the triangle AEC, right-angled at A, the side AC and all the angles, to find the hypotenuse CE, which is the distance of the moon from the center of the earth.

REFRACTION.

88. While parallax depresses the celestial bodies subject to it, *refraction elevates them;* and it affects alike the most distant as well as nearer bodies, being occasioned by the change of direction which light undergoes in passing through the atmosphere. Let us conceive of the atmosphere as made up of a great number of concentric strata, as AA, BB, CC, and DD (Fig. 8), increasing rapidly in density (as is known to be the

Fig. 8.

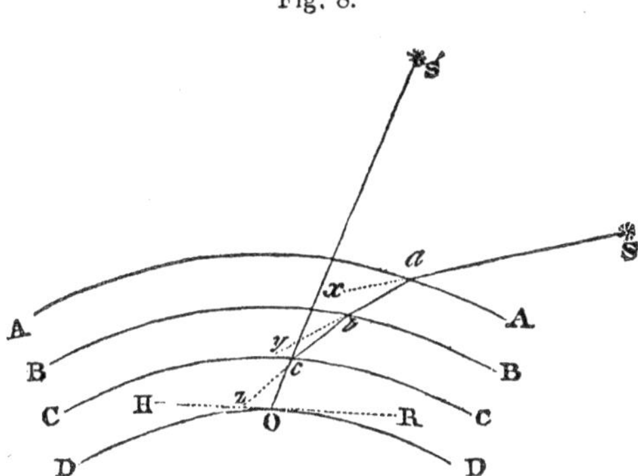

fact) in approaching near to the surface of the earth. Let S be a star, from which a ray of light S*a* enters the atmosphere at *a*, where, being turned towards the radius of the convex surface, it would change its direction into the line *ab*, and again into *bc*, and *c*O, reaching the eye at O. Now, since an object always appears in the direction in which the light finally strikes the eye, the star would be seen in the direction of the last ray *c*O, and the star would apparently change its place, in consequence of refraction, from S to S′, being elevated out of its true position. Moreover, since on account of the constant increase of density in descending through the atmosphere, the light would be continually turned out of its course more and more; it would therefore move, not in the polygon represented in the figure, but in a corresponding curve, whose curvature is rapidly increased near the surface of the earth.

89. When a body is in the zenith, since a ray of light from it enters the atmosphere at right angles to the refracting medium, it suffers no refraction. Consequently, the position of the heavenly bodies, when in the zenith, is not changed by refraction; while, near the horizon, where a ray of light strikes the medium very obliquely, and traverses the atmosphere through its densest part, the refraction is greatest. The following numbers, taken at different altitudes, will show how rapidly refraction diminishes from the horizon upward. The amount of refraction at the horizon is 34′ 00″. At different elevations it is as follows.

Elevation.	Refraction.	Elevation.	Refraction.
0° 10′	32′ 00″	30°	1′ 40″
0 20	30 00	40	1 09
1 00	24 25	45	0 58
5 00	10 00	60	0 33
10 00	5 20	80	0 10
20 00	2 39	90	0 00

From this table it appears, that while refraction at the horizon is 34 minutes, at so small an elevation as only 10 minutes above the horizon it loses 2 minutes, more than the entire change from the elevation of 30° to the zenith. From the horizon to 1° above, the refraction is diminished nearly 10 minutes. The amount at the horizon, at 45°, and at 90° respectively, is 34′, 58″, and 0. In finding the altitude of a heavenly body, the effect of parallax must be added, but that of refraction subtracted.

90. Let us now learn the method by which the amount of refraction at different elevations is ascertained. To take the simplest case, we will suppose ourselves in a high latitude, where some of the stars within the circle of perpetual apparition pass through the zenith of the place. We measure the distance of such a star from the pole when on the meridian above the pole, that is, in the zenith, where it is not at all affected by refraction, and again its distance from the pole in its lower culmination. Were it not for refraction, these two polar distances would be equal, since, in the diurnal revolution of a star, it is in fact always at the same distance from the pole; but, on account of refraction, the lower distance will be less than the upper, and the difference between the two will show the amount of refraction at the lower culmination, the latitude of the place being known.

Example. At Paris, latitude 48° 50′, a star was observed to pass the meridian 6′ north of the zenith, and consequently

41° 4′ from the pole.* It should have passed the meridian at the same distance below the pole, but the distance was found to be only 40° 57′ 35″. Hence, 41° 4′—40° 57′ 35″=6′ 25″ is the refraction due to that altitude, that is, at the altitude of 7° 46′=(48° 50′—41° 4′). By taking similar observations in various places situated in high latitudes, the amount of refraction might be ascertained for a number of different altitudes, and thus the law of increase in refraction, as we proceed from the zenith towards the horizon, might be ascertained.

91. Another method of finding the refraction at different altitudes, is as follows. Take the altitude of the sun or a star, whose right ascension and declination are known, and note the time by the clock. Observe also when it crosses the meridian, and the difference of time between the two observations will give the hour angle ZPx (Fig. 9). In this triangle ZPx we

Fig. 9.

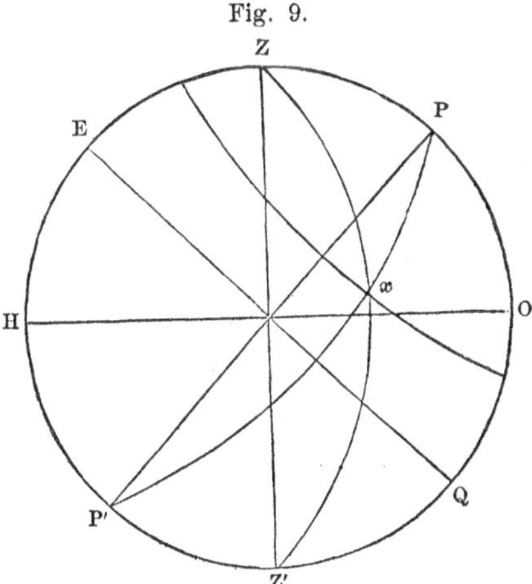

also know PZ the co-latitude and Px the co-declination. Hence we can find the co-altitude Zx, and of course the true altitude. Compare the altitude thus found with that before determined by observation, and the difference will be the refraction due to the apparent altitude.

Ex. On May 1, 1738, at 5h. 20m. in the morning, Cassini

* For the polar distance of the place=90—48° 50′=41° 10′; and 41° 10—6′ =41° 4′.

observed the altitude of the sun's center at Paris to be 5° 0′ 14″. The latitude of Paris being 48° 50′ 10″, and the sun's declination at that time being 15° 0′ 25″: *Required the refraction.*

By spherical trigonometry, Zx will be found equal to 85° 10′ 8″; consequently, the true altitude was 4° 49′ 52″. Now to 5° 0′ 14″, the apparent altitude, 9″ must be added for parallax, and we have 5° 0′ 23″, the apparent altitude corrected for parallax. Hence, 5° 0′ 23″−4° 49′ 52″=10′ 31″, the refraction at the apparent altitude 5° 0′ 14″.*

92. By these and similar methods, we could easily determine the refraction due to any elevation above the horizon, provided the refracting medium (the atmosphere) were always uniform. But this is not the fact: the refracting power of the atmosphere is altered by changes in density and temperature.† Hence, in delicate observations, it is necessary to take into the account the state of the barometer and of the thermometer, the influence of the variations of each having been very carefully investigated, and rules having been given accordingly. With every precaution to insure accuracy, on account of the variable character of the refracting medium, the tables are not considered as entirely accurate to a greater distance from the zenith than 74°; but almost all astronomical observations are made at a greater altitude than this.

93. Since the whole amount of refraction near the horizon exceeds 33′, and the diameters of the sun and moon are severally less than this, these luminaries are in view both before they have actually risen and after they have set.

94. The rapid increase of refraction near the horizon is strikingly evinced by the *oval* figure which the sun assumes when near the horizon, and which is seen to the greatest advantage when light clouds enable us to view the solar disk. Were all parts of the sun equally raised by refraction, there

---

* Gregory's Ast., p. 65.

† It is said that the effects of *humidity* are insensible; for the most accurate experiments seem to prove that watery vapor diminishes the density of air in the same ratio as its own refractive power is greater than that of air. (New Encyc. Brit., iii., 762.)

would be no change of figure; but since the lower side is more refracted than the upper, the effect is to shorten the vertical diameter, and thus to give the disk an oval form. This effect is particularly remarkable when the sun, at its rising or setting, is observed from the top of a mountain, or at an elevation near the sea-shore; for in such situations the rays of light make a greater angle than ordinary with a perpendicular to the refracting medium, and the amount of refraction is proportionally greater. In some cases of this kind, the shortening of the vertical diameter of the sun has been observed to amount to 6′, or about one-fifth of the whole.*

95. The apparent *enlargement of the sun and moon in the horizon*, arises from an optical illusion. These bodies in fact are not seen under so great an angle when in the horizon, as when on the meridian, for they are nearer to us in the latter case than in the former. The distance of the sun is indeed so great that it makes very little difference in his apparent diameter, whether he is viewed in the horizon or on the meridian; but with the moon the case is otherwise; its angular diameter, when measured with instruments, is perceptibly larger at the time of its culmination. Why then do the sun and moon appear so much larger when near the horizon? It is owing to that general law, explained in optics, by which we judge of the magnitudes of distant objects, not merely by the angle they subtend at the eye, but also by our impressions respecting their distance, allowing, under a given angle, a greater magnitude as we imagine the distance of a body to be greater. Now, on account of the numerous objects usually in sight between us and the sun, when on the horizon, he appears much further removed from us than when on the meridian, and we assign to him a proportionally greater magnitude. If we view the sun, in the two positions, through smoked glass, no such difference of size is observed, for here no objects are seen but the sun himself.

---

* In extreme cold weather, this shortening of the sun's vertical diameter sometimes exceeds this amount.

TWILIGHT.

96. *Twilight* also is another phenomenon depending upon the agency of the earth's atmosphere. It is due partly to refraction and partly to reflection, but mostly the latter. While the sun is within 18° of the horizon, before it rises or after it sets, some portion of its light is conveyed to us by means of numerous reflections from the atmosphere. Let AB (Fig. 10) be

Fig. 10.

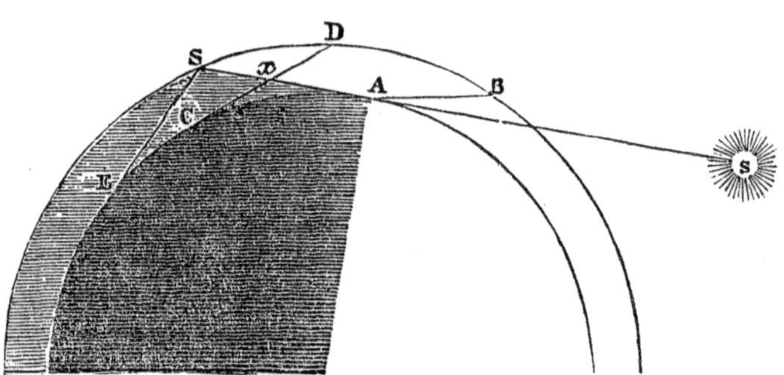

the horizon of the spectator at A, and let SS be a ray of light from the sun when it is two or three degrees below the horizon. Then to the observer at A, the segment of the atmosphere ABS would be illuminated. To a spectator at C, whose horizon was CD, the small segment SDx would be the twilight; while, at E, the twilight would disappear altogether.

97. At the equator, where the circles of daily motion are perpendicular to the horizon, the sun descends through 18° in an hour and twelve minutes ($\frac{18}{15}=1\frac{1}{5}$h.), and the light of day therefore declines rapidly, and as rapidly advances after daybreak in the morning. At the pole, a constant twilight is enjoyed while the sun is within 18° of the horizon, occupying nearly two-thirds of the half year when the direct light of the sun is withdrawn, so that the progress from continual day to constant night is exceedingly gradual. To the inhabitants of an oblique sphere, the twilight is longer in proportion as the place is nearer the elevated pole.

98. Were it not for the power the atmosphere has of dispersing the solar light, and scattering it in various directions, no objects would be visible to us out of direct sunshine; every shadow of a passing cloud would be pitchy darkness; the stars would be visible all day, and every apartment into which the sun had not direct admission, would be involved in the obscurity of night. This scattering action of the atmosphere on the solar light, is greatly increased by the irregularity of temperature caused by the sun, which throws the atmosphere into a constant state of undulation, and by thus bringing together masses of air of different temperatures, produces partial reflections and refractions at their common boundaries, by which means much light is turned aside from the direct course, and diverted to the purposes of general illumination.* In the upper regions of the atmosphere, as on the tops of very high mountains, where the air is too much rarefied to reflect much light, the sky assumes a black appearance, and stars become visible in the day-time.

CHAPTER IV.

OF TIME.

99. Time *is a measured portion of indefinite duration.*

Any event may be taken as a measure of time, which divides a portion of duration into equal parts; as the pulsations of the wrist, the vibrations of a pendulum, or the passage of sand from one vessel into another, as in the hour-glass.

100. The great standard of time is the period of the revolution of the earth on its axis, which, by the most exact observations, is found to be always the same. The time of the earth's revolution on its axis is called a *sidereal day*, and is determined by the revolution of a star from the instant it crosses the meridian, until it comes round to the meridian again. This interval being called a sidereal day, it is divided into 24 *sidereal hours.* Observations taken upon numerous stars, in different

* Herschel.

ages of the world, show that they all perform their diurnal revolutions in the same time, and that their motion during any part of the revolution is perfectly uniform.

101. *Solar time* is reckoned by the apparent revolution of the sun, from the meridian round to the same meridian again. Were the sun stationary in the heavens, like a fixed star, the time of its apparent revolution would be equal to the revolution of the earth on its axis, and the solar and the sidereal days would be equal. But since the sun passes from west to east, through 360° in 365¼ days, it moves eastward nearly 1° a day (59′ 8″.3). While, therefore, the earth is turning round on its axis, the sun is moving in the same direction, so that when we have come round under the same celestial meridian from which we started, we do not find the sun there, but he has moved eastward nearly a degree, and the earth must perform so much more than one complete revolution, in order to come under the sun again. Now since a place on the earth gains 360° in 24 hours, how long will it take to gain 1°?

$$360 : 24 :: 1 : \frac{24}{360} = 4^{m} \text{ nearly.}$$

Hence the solar day is about 4 minutes longer than the sidereal; and if we were to reckon the sidereal day 24 hours, we should reckon the solar day 24h. 4m. To suit the purposes of society at large, however, it is found most convenient to reckon the solar day 24 hours, and to throw the fraction into the sidereal day. Then,

24h. 4m. : 24 :: 24 : 23h. 56m. (23h. 56^{m} 4^{s}.09)=the length of a sidereal day.

102. The solar days, however, do not always differ from the sidereal by precisely the same fraction, since the increments of right ascension (Art. 37), which measure the difference between a sidereal and a solar day, are not equal to each other. *Apparent time* is time reckoned by the revolutions of the sun from the meridian to the meridian again. These intervals being unequal, of course the apparent solar days are unequal to each other.

103. *Mean time* is time reckoned by the *average* length of all the solar days throughout the year. This is the period which constitutes the civil day of 24 hours, beginning when the sun is on the lower meridian, namely, at 12 o'clock at night, and counted by 12 hours from the lower to the upper culmination, and from the upper to the lower. The *astronomical* day is the apparent solar day counted through the whole 24 hours, instead of by periods of 12 hours each, and begins at noon. Thus 10th day and 14th hour of astronomical time, would be 11th day and 2d hour of civil time.

104. Clocks are usually regulated so as to indicate mean solar time; yet as this is an artificial period, not marked off, like the sidereal day, by any natural event, it is necessary to know how much is to be added to or subtracted from the apparent solar time, in order to give the corresponding mean time. The interval by which apparent time differs from mean time, is called the *equation of time.* If a clock were constructed (as it may be) so as to keep exactly with the sun, going faster or slower according as the increments of right ascension were greater or smaller, and another clock were regulated to mean time, then the difference of the two clocks at any period, would be the equation of time for that moment. If the apparent clock were *faster* than the mean, then the equation of time must be subtracted; but if the apparent clock were *slower* than the mean, then the equation of time must be added, to give the mean time. The two clocks would differ most about the 3d of November, when the apparent time is $16\frac{1}{4}^{m}$ greater than the mean (16^{m} 17^{s}). But, since apparent time is sometimes greater and sometimes less than mean time, the two must obviously be sometimes equal to each other. This is in fact the case four times a year, namely, April 15th, June 15th, September 1st, and December 22d. These epochs, however, do not remain constant; for, on account of the change in the position of the perihelion, or the point where the earth is nearest the sun (which shifts its place from west to east about 12″ a year), the period when the sun's motions are most rapid, as well as that when they are slowest, will occur at different parts of the year. The change is indeed exceedingly small in a single year; but in the progress of ages, the time of year when the sun's motion in its orbit is most accelerated, will not be, as at present, on the first of January, but may fall on the first of March, June, or any other day of the year, and the amount of the equation of time is obviously affected by the sun's distance from its perihelion, since the sun moves most rapidly when nearest the perihelion, and slowest when furthest from that point.

105. *The inequality of the solar days depends on two causes, the unequal motion of the earth in its orbit, and the inclination of the equator to the ecliptic.*

First, on account of the eccentricity* of the earth's orbit, the earth actually moves faster from the autumnal to the vernal equinox, than from the vernal to the autumnal, the difference of the two periods being about eight days (7d. 17h. 17m.)

Fig. 11.

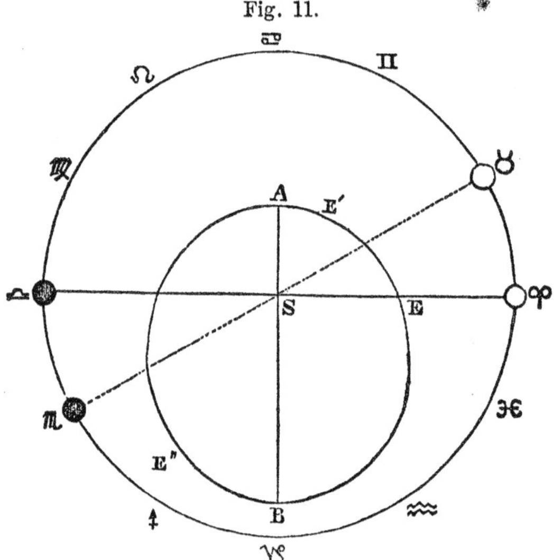

Thus, let AEB (Fig. 11) represent the earth's orbit, S being the place of the sun, A the *perihelion*, or nearest distance of the earth from the sun, B the *aphelion*, or greatest distance, and E, E′, E″, positions of the earth in different points of its orbit. The place of the earth among the signs is the part of

* The exact figure of the earth's orbit will be more particularly shown hereafter. All that the student requires to know, in order to understand the present subject, is, that the earth's orbit is an ellipse, and that the earth's *real* motion, and consequently the sun's *apparent* motion, is greater in proportion as the earth is nearer the sun.

the heavens to which it would be referred if seen from the sun; and the place of the sun is the part of the heavens to which it is referred as seen from the earth. Thus, when the earth is at E, it is said to be in Aries; and as it moves from E through E′ to A, its path in the heavens is through Aries, Taurus, Gemini, &c. Meanwhile the sun takes its place successively in Libra, Scorpio, Sagittarius, &c. Now, as will be shown more fully hereafter, the earth moves faster when proceeding from Aries through its perihelion to Libra, than from Libra through its aphelion to Aries, and, consequently, describes the half of its apparent orbit in the heavens, ♈, ♋, ♎, sooner than the half ♎, ♑, ♈. The line of the apsides, that is, the major axis of the ellipse, is so situated at present, that the perihelion is in the sign Cancer, nearly 100° (99° 30′ 5″) from the vernal equinox. The earth passes through it about the first of January, and then its velocity is the greatest in the whole year, being always greater as the distance is less, the angular velocity being inversely as the square of the distance, as will be shown by and by.

106. But differences of time are not reckoned on the ecliptic, but on the equinoctial; for the ecliptic being oblique to the meridian in the diurnal motion, and cutting it at different angles at different times, equal portions will not pass under the meridian in equal times, and therefore such portions could not be employed, as they are in the equinoctial, as measures of time. If, therefore, the sun moved uniformly in his orbit, so as to make the daily increments of longitude equal, still the corresponding arcs of right ascension, which determine the lengths of the solar day, would be unequal. Let us start from the equinox, from which both longitude and right ascension are reckoned, the former on the ecliptic, the latter on the equinoctial. Suppose the sun has described 70° of longitude; then to ascertain the corresponding arc of right ascension, we let a meridian pass through the sun: the point where it cuts the equator gives the sun's right ascension. Now since the ecliptic makes an acute angle with the meridian, while the equinoctial makes a right angle with it, consequently the arc of longitude is greater than the arc of right ascension. The difference, however, grows constantly less and less as we approach the tropic, as the angle made between the ecliptic and the meridian constantly increases, until, when we reach the tropic, the meridian is at right angles to both circles, and the longitude and right ascension each equals 90°, and they are of course equal to each other. Beyond this, from the tropic to the other equinox, the arc of the ecliptic intercepted between the meridian and the autumnal equinox being greater than the corresponding arc of the equinoctial, of course its supplement, which measures the longitude, is less than the supplement of the corresponding arc of the equator which measures the right ascension. At the autumnal equinox again, the right ascension and longitude become equal. In a similar manner we might show that the *daily increments* of longitude and right ascension are unequal.

In order to illustrate the foregoing points, let ♈ ♎ (Fig. 12)

Fig. 12.

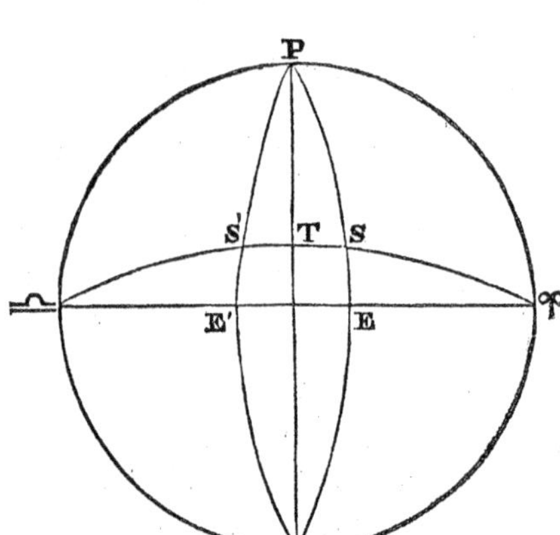

represent the equator, ♈ T ♎ the ecliptic, and PSE, PS′E′, two meridians meeting the sun in S and S′. Then in the triangle ♈ES, the arc of longitude ♈S, is greater than ♈E, the corresponding arc of right ascension; but towards the tropic the difference between the two arcs evidently grows less and less, until at T the arcs become equal, being each 90°. But, beyond the tropic, since ♈E′♎, ♈S′♎, are equal to each other, each being equal to 180°, and since S′♎ is greater than E′♎, therefore ♈S′ must be less than ♈E′.

107. As the *whole arc* of right ascension reckoned from the first of Aries, does not keep uniform pace with the corresponding arc of longitude, so the *daily increments* of right ascension differ from those of longitude. If we suppose in the quadrant ♈T, points taken to mark the progress of the sun from day to day, and let meridians like PSE pass through these points, the arcs of the ecliptic embraced between the meridians will be the daily increments of longitude, while the corresponding parts of the equinoctial will be the daily increments of right ascension. Near ♈, the oblique direction in which the ecliptic cuts the meridian, will make the daily increments of longitude exceed those of right ascension; but this advantage is diminished as we approach the tropic, where the ecliptic becomes less oblique, and finally parallel to the equinoctial; while the convergence of the meridians contributes still further to lessen the ratios of the daily increments of longitude to those of right ascension. Hence, at first, the diurnal arcs of right ascension are less than those of longitude, but afterward greater; and they continue greater for a similar distance beyond the tropic.

108. From the foregoing considerations it appears, that the diurnal arcs of right ascension, by which the difference between the sidereal and the solar days is measured, are unequal, on account both of the unequal motion of the sun in his orbit, and of the inclination of his orbit to the equinoctial.

109. As astronomical time commences when the *sun* is on the meridian, so sidereal time commences when the vernal equinox is on the meridian, and is also counted from 0 to 24 hours. By 3 o'clock, for instance, of sidereal time, we mean that it is three hours since the *vernal equinox* crossed the meridian; as we say it is 3 o'clock of astronomical or of civil time, when it is three hours since the *sun* crossed the meridian.

THE CALENDAR.

110. The *astronomical year* is the time in which the sun makes one revolution in the ecliptic, and consists of 365d. 5h. 48m. $51^s.60$. The *civil year* consists of 365 days. The difference is nearly 6 hours, making one day in four years.

111. The most ancient nations determined the number of days in the year by means of the *stylus*, a perpendicular rod which cast its shadow on a smooth plane, bearing a meridian line. The time when the shadow was shortest, would indicate the day of the summer solstice; and the number of days which elapsed until the shadow returned to the same length again, would show the number of days in the year. This was found to be 365 whole days, and accordingly this period was adopted for the civil year. Such a difference, however, between the civil and astronomical years, at length threw all dates into confusion. For, if at first the summer solstice happened on the 21st of June, at the end of four years, the sun would not have reached the solstice until the 22d of June, that is, it would have been behind its time. At the end of the next four years the solstice would fall on the 23d; and in process of time it would fall successively on every day of the year. The same would be true of any other fixed date. Julius Cæsar made the first correction of the calendar, by introducing an intercalary day every fourth year, making February to consist of 29 instead of 28 days, and of course the whole year to consist of 366 days. This fourth year was denominated *Bissextile.** It is also called Leap Year.

112. But the true correction was not 6 hours, but 5h. 49m.; hence the intercalation was too great by 11 minutes. This small fraction would amount in 100 years to ¾ of a day, and in 1000 years to more than 7 days. From the year 325 to 1582, it had in fact amounted to about 10 days; for it was known that in 325, the vernal equinox fell on the 21st of March, whereas, in 1582 it fell on the 11th. In order to restore the equinox to the same date, Pope Gregory XIII. decreed that the year should be brought forward ten days, by reckoning the 5th of October the 15th. In order to prevent the calendar from falling into confusion afterward, the following rule was adopted:

Every year whose number is not divisible by 4 *without a remainder, consists of* 365 *days; every year which is so divisible, but is not divisible by* 100, *of* 366; *every year divisible by* 100

---

* The sextus dies ante Kalendas being reckoned twice (Bis).

but not by 400, *again of* 365; *and every year divisible by* 400, *of* 366.

Thus the year 1838, not being divisible by four, contains 365 days, while 1836 and 1840 are leap years. Yet to make every fourth year consist of 366 days would increase it too much by about ¾ of a day in 100 years; therefore every hundredth year has only 365 days. Thus 1800, although divisible by 4, was not a leap year, but a common year. But we have allowed a *whole* day in a hundred years, whereas we ought to have allowed only *three-fourths* of a day. Hence, in 400 years we should allow a day too much, and therefore we let the 400th year remain a leap year. This rule involves an error of less than a day in 4237 years.* If the rule were extended by making every year divisible by 4,000 (which would now consist of 366 days) to consist of 365 days, the error would not be more than one day in 100,000 years.†

113. This reformation of the calendar was not adopted in England until 1752, by which time the error in the Julian calendar amounted to about 11 days. The year was brought forward, by reckoning the 3d of September the 14th. Previous to that time the year began the 25th of March; but it was now made to begin on the 1st of January, thus shortening the preceding year, 1751, one quarter.‡

114. As in the year 1582, the error in the Julian calendar amounted to 10 days, and increased by ¾ of a day in a century, at present the correction is 12 days; and the number of the year will differ by one with respect to dates between the 1st of January and the 25th of March.

Examples. General Washington was born Feb. 11, 1731, old style; to what date does this correspond in new style?

As the date is the earlier part of the 18th century, the correction is 11 days, which makes the birthday fall on the 22d

* Woodhouse, p. 874.

† Herschel's Ast., p. 384.

‡ Russia, and the Greek Church generally, adhere to the old style. In order to make the Russian dates correspond to ours, we must add to them 12 days. France and other Catholic countries, adopted the Gregorian calendar soon after it was promulgated.

of February; and since the year 1731 closed the 25th of March, while according to new style 1732 would have commenced on the preceding 1st of January; therefore, the time required is Feb. 22d, 1732. It is usual, in such cases, to write both years, thus: Feb. 11, 1731–2, O. S.

2. A great eclipse of the sun happened May 15th, 1836; to what date would this time correspond in old style?

Ans. May 3d.

115. *The common year begins and ends on the same day of the week; but leap year ends one day later in the week than it began.*

For 52×7=364 days; if, therefore, the year begins on Tuesday, for example, 364 days would complete 52 weeks, and one day would be left to begin another week, and the following year would begin on Wednesday. Hence, generally, any day of the month is one day later in the week than the same day of the preceding year. Thus, July 4th, 1861, falls on Thursday; 1862, on Friday; 1863, on Saturday. But, in leap-year, this rule applies only till the end of February. From that time to the same date in the year following, every day of a month falls *two* days later in the week than during the previous year. Thus, July 4th, 1871, is Tuesday; 1872, Thursday; and February 2d, 1872, is Friday; 1873, Sunday.

116. Fortunately for astronomy, the confusion of dates involved in different calendars affects recorded observations but little. Remarkable eclipses, for example, can be calculated back for several thousand years, without any danger of mistaking the day of their occurrence; and whenever any such eclipse is so interwoven with the account given by an ancient author of some historical event, as to indicate precisely the interval of time between the eclipse and the event, and at the same time completely to identify the eclipse, that date is recovered and fixed forever.*

* An elaborate view of the Calendar may be found in Delambre's Astronomy, t. III. A useful table for finding the day of the week of any given date, is inserted in the American Almanac for 1832, p. 72.

CHAPTER V.

OF ASTRONOMICAL INSTRUMENTS AND PROBLEMS—FIGURE AND DENSITY OF THE EARTH.

117. THE most ancient astronomers employed no instruments for measuring angles, but acquired their knowledge of the heavenly bodies by long-continued and most attentive inspection with the naked eye. In the Alexandrian school, about 300 years before the Christian era, instruments began to be freely used, and thenceforward trigonometry lent a powerful aid to the science of astronomy. Tycho Brahe, in the 16th century, formed a new era in practical astronomy, and carried the measurement of angles to 10″,—a degree of accuracy truly wonderful, considering that he had not the advantage of the telescope. By the application of the telescope to astronomical instruments, a far better defined view of objects was acquired, and a far greater degree of refinement was attainable. The astronomers royal of Great Britain perfected the art of observation, bringing the measurement of angles to 1″, and the estimation of differences of time to $\frac{1}{10}$ of a second. Beyond this degree of refinement it is supposed that we cannot advance, since unavoidable errors arising from the uncertainties of refraction, and the necessary imperfection of instruments, forbid us to hope for a more accurate determination than this. But a little reflection will show us, that 1″ on the limb of an astronomical instrument, must be a space exceedingly small. Suppose the circle, on which the angle is measured, be one foot in diameter. Then $\frac{12\times 3.14159}{360}=\frac{1}{10}$ inch= space occupied by 1°. Hence $\frac{1}{10\times 60}=\frac{1}{600}$=space of 1′, and $\frac{1}{36000}$=space of 1″. Such minute angles can be measured only by large circles. If, for example, a circle is 20 feet in diameter, a degree on its periphery would occupy a space 20 times as large as a degree on a circle of 1 foot. A degree therefore of the limb of such an instrument would occupy a space of 2 inches; one minute, $\frac{1}{30}$ of an inch; and one second, $\frac{1}{1800}$ of an inch.

118. But the actual divisions on the limb of an astronomical instrument never extend to seconds: in the smaller instruments they reach only to 10′, and on the largest rarely lower than 1′. The subdivision of these spaces is carried on by means of the Vernier, which may be thus defined:

A VERNIER *is a contrivance attached to the graduated limb of an instrument, for the purpose of measuring aliquot parts of the smallest spaces into which the instrument is divided.*

The vernier is usually a narrow zone of metal, which is made to slide on the graduated limb. Its divisions correspond to those on the limb, except that they are a little larger,* one-tenth, for example, so that ten divisions on the vernier would equal eleven on the limb. Suppose now that our instrument is graduated to degrees only, but the altitude of a certain star is found to be 40° and *a fraction*, or $40°+x$. In order to estimate the amount of this fraction, we bring the zero-point of the vernier to coincide with the point which indicates the exact altitude, or $40°+x$. We then look along the vernier until we find where one of its divisions coincides with one of the divisions of the limb. Let this be at the fourth division of the vernier. In four divisions, therefore, the vernier has gained upon the divisions of the limb, a space equal to x; and since, in the case supposed, it gains $\frac{1}{10}$ of a degree, or 6′ at each division, the entire gain is 24′, and the arc in question is 40° 24′.

119. As the vernier is used in the barometer, where its application is more easily seen than in astronomical instruments, while the principle is the same in both cases, let us see how it is applied to measure the exact height of a column of mercury. Let AB (Fig. 13) represent the upper part of a barometer, the level of the mercury being at C, namely, at 30.3 inches, and nearly another tenth. The vernier being brought

* In the more modern instruments the divisions of the vernier are smaller than those of the limb.

(by a screw which is usually attached to it) to coincide with the surface of the mercury, we look along down the scale, until we find that the coincidence is at the 8th division of the vernier. Now as the vernier gains $\frac{1}{10}$ of $\frac{1}{10} = \frac{1}{100}$ of an inch at each division upward, it of course gains $\frac{8}{100}$ in eight divisions. The fractional quantity, therefore, is .08 of an inch, and the height of the mercury is 30.38. If the divisions of the vernier were such, that each gained $\frac{1}{60}$ (when 60 on the vernier would equal 61 on the limb) on a limb divided into degrees, we could at once take off minutes; and were the limb graduated to minutes, we could in a similar way read off seconds.

Fig. 13.

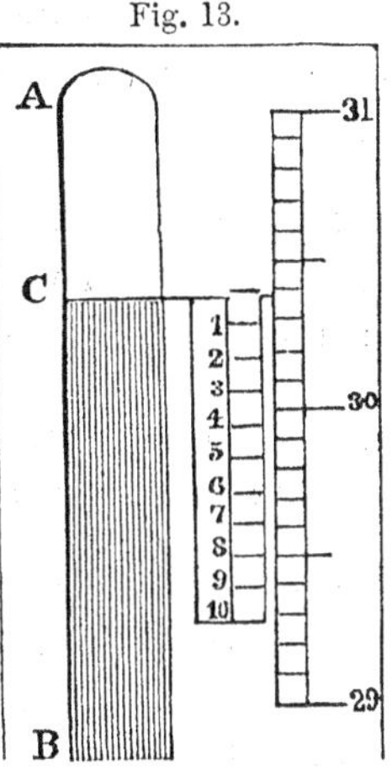

120 The instruments most used for astronomical observations, are the Transit Instrument, the Astronomical Clock, the Mural Circle, and the Sextant. A large portion of all the observations made in an astronomical observatory, are taken *on the meridian*. When a heavenly body is on the meridian, being at its highest point above the horizon, it is then least affected by refraction and parallax; its zenith distance (from which its altitude and declination are easily derived) is readily estimated; and its right ascension may be very conveniently and accurately determined by means of the astronomical clock. Having the right ascension and declination of a heavenly body, various other particulars respecting its position may be found, as we shall see hereafter, by the aid of spherical trigonometry. Let us then first turn our attention to the instruments employed for determining the right ascension and declination. They are the Transit Instrument, the Astronomical Clock, and the Mural Circle.

121. The *Transit Instrument* is a telescope, which is fixed permanently in the meridian, and moves only in that plane. It rests on a horizontal axis, which consists of two hollow cones applied base to base, a form uniting lightness and strength. The two ends of the axis rest on two firm supports, as pillars of stone, for example, usually built up from the ground, and so related to the building as to be as free as possible from all agitation. In figure 14, AD represents the telescope, E, W, massive stone pillars supporting the horizontal axis, beneath which is seen a spirit-level (which is used to bring the axis to a horizontal position), and *n* a vertical circle graduated into degrees and minutes. This circle serves the purpose of placing the instrument at any required altitude or distance from the zenith, and of course for determining altitudes and zenith distances.

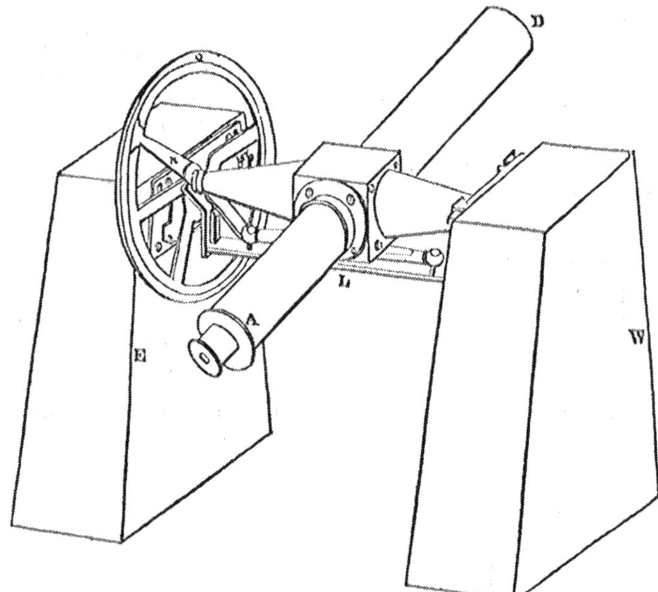

122. Various methods are described in works on practical astronomy, for placing the Transit Instrument accurately in the meridian. The following method, by observations on the pole-star, may serve as an example. If the instrument be directed towards the north star, and so adjusted that the star Alioth (the first in the tail of the Great Bear) and the pole-star are both in the same vertical circle, the former below the pole and the latter above it, the instrument will be *nearly* in the plane of the meridian. To adjust it more exactly, compare the time occupied by the pole-star in passing from its upper to its lower culmination, with the time of passing from its lower to its upper culmination. These two intervals ought to be precisely equal; and if they are so, the instrument is truly placed in the meridian; but if they are not equal, the position of the instrument must be shifted until they become exactly equal.

123. The *line of collimation* of a telescope, is a line joining the center of the object-glass with the center of the eye-glass. When the transit instrument is properly adjusted, this line, as the instrument is turned on its axis, moves in the plane of the meridian. Having, by means of the vertical circle *n*, set the instrument at the known altitude or zenith distance of any star, upon which we wish to make observations, we wait until the star enters the field of the telescope, and note the exact instant when it crosses the vertical wire in the center of the field, which wire marks the true plane of the meridian. Usually, however, there are placed in the focus of the eye-glass five parallel wires or threads, two on each side of the central wire, and all at equal distances from each other, as is represented in the following diagram. The time of arriving at each of the wires being noted, and all the times added together and divided by the number of observations, the result gives the instant of crossing the central wire.

Fig. 15.

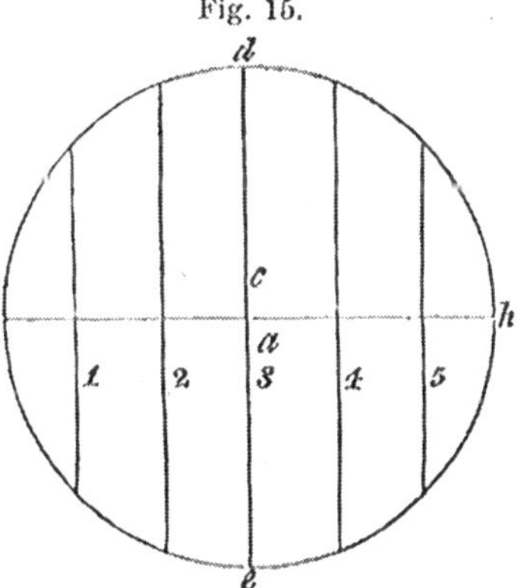

124. The *astronomical clock* is the constant companion of the transit instrument. It is regulated to keep exact pace with the stars, and of course with the earth's diurnal rotation; that is, it denotes *sidereal time*, measuring off one hour for every 15° of diurnal motion in a star. The sidereal day begins at the moment when the vernal equinox crosses the meridian; but as the culmination of the equinox occurs about 4 minutes later from day to day through the year, the sidereal time may differ from the solar time by any quantity whatever. The sidereal clock may point to 3h. 20m., in the morning, at noon, or any other time of day, because it merely shows that 3h. 20m. have elapsed since the equinox was on the meridian. Hence, when a star is on the meridian, the sidereal clock shows its right ascension.

An astronomical clock must have a compensation pendulum, and be made as perfect as possible. Its uniformity of movement can be tested by the transit instrument, and a list of right ascensions of stars. It is not so important that it should point to 0h. 0m. 0s. when the equinox is on the meridian, or that it should not *gain* or *lose* compared with the revolution of the stars, as that it should move *uniformly* through the day, and from day to day. It is not customary, therefore, to alter the clock, after it is once set, but to note from day to day how much it is out of the way, and how fast it gains or loses. The first is called the *error*, the last the *rate*. If these are known, then the exact time of an observation can be obtained.

125. To observe the transit of a star, the *eye* must discern the instant of its bisection by the wire, and the *ear* attend to the beat of the clock, the seconds being counted from the last completed minute before the observation began. If the bisection occurs between two beats, as it commonly does, the observer needs much practice to be able to divide the second accurately into tenths, and decide at which of them the transit takes place. What is now known as the *American method* of observing transits, and recording them by electro-magnetism, gives great facility and accuracy to this most difficult and important part of the work.

The pendulum of the observatory clock is arranged to close the circuit of a battery, and break it again at the beginning of every beat. The closing of the circuit gives a small lateral motion to the registering pen, under which the paper is advancing about an inch per second. Thus the seconds are all permanently recorded by notches one inch asunder in a straight

Fig. 15′.

a b A c d

line, as *a*, *b*, *c*, &c. (Fig. 15′). The mark at the beginning of each *minute* has some peculiarity by which it may be distinguished from the rest. The observer has under his hand a key, which by a quick touch will also close and break the circuit. Whenever a star is on one of the wires of the transit telescope, he touches the key, the pen is moved aside and indents the line, as at A, and the observation is thus recorded; and the place where this motion commenced between the second-marks can afterwards be carefully examined. Thus, without the distraction of attending to the clock, he can record the transits of all the wires; and if he only notices within what minute the work begins, he can read the entire record with accuracy to the $\frac{1}{10}$ or even the $\frac{1}{100}$ of a second. Since the general adoption of this method, the number of wires in the focus of the eye-glass has been increased from 5 to 30 or 40, in order to secure a more perfect result.

126. The vertical circle (n, Fig. 14), usually connected with the Transit Instrument, affords the means of measuring arcs on the meridian, as meridian altitudes, zenith distances, and declinations; but as the circle must necessarily be small, and therefore incapable of measuring very minute angles, the *Mural Circle* is usually employed for measuring arcs of the meridian. The Mural Circle is a graduated circle, usually of very large size, fixed permanently in the plane of the meridian, and attached firmly to a perpendicular *wall*. It is made of large size, sometimes 11 feet in diameter, in order that very small angles may be measured on its limb; and it is attached to a massive wall of solid masonry in order to insure perfect steadiness, a point the more difficult to attain in proportion as the instrument is heavier. The annexed diagram represents a Mural Circle fixed to its wall and ready for observations. It will be seen that every expedient is employed to give the instrument firmness of parts and steadiness of position. Its radii are composed of hollow cones, uniting lightness and strength, and its telescope revolves on a large horizontal axis, fixed as firmly as possible in a solid wall. The graduations are made on the outer rim of the instrument, and are read off by six microscopes (called *reading microscopes*), attached to the wall, one of which is represented at A, and the places of the five others are marked by the letters B, C, D, E, F. Six are used, in order that by taking the mean of such a number of readings, a higher degree of accuracy may be insured, than could be obtained by a

Fig. 16.

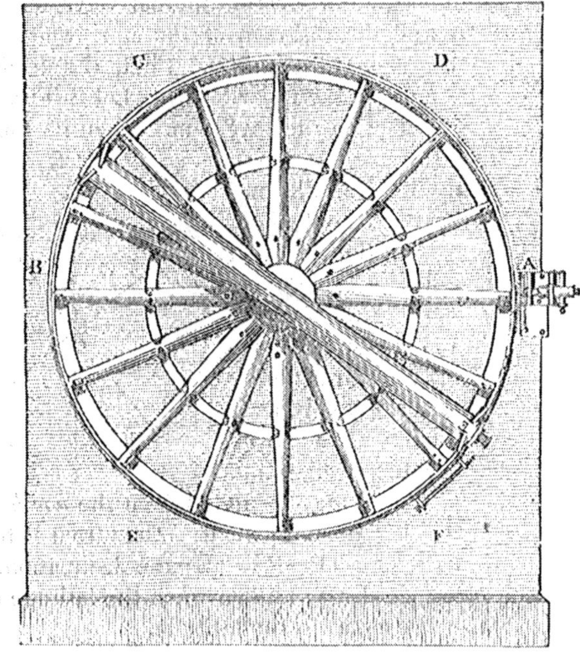

single reading. In a circle of six feet diameter, like that represented in the figure, the divisions may be easily carried to five minutes each. The microscope (which is of the variety called *compound microscope*) forms an enlarged image of each of these divisions in the focus of the eye-glass. With it is combined the principle of the *micrometer.* This is effected by placing in the focus a delicate wire, which may be moved by means of a screw in a direction parallel to the divisions of the limb, and which is so adjusted to the screw as to move over the whole magnified space of five minutes by five revolutions of the screw. Of course one revolution of the screw measures one minute. Moreover, if the screw itself is made to carry an index attached to its axis and revolving with it over a disk graduated into sixty equal parts, then the space measured by moving the index over one of these parts will be one second.

127. We have before shown (Art. 124), the method of finding the right ascension of a star by means of the Transit Instrument and the clock. The declination may be obtained by means of the mural circle in several different ways, our object being always to find the distance of the star, when on the meridian, from the equator (Art. 37). First, the declination may be found from the *meridian altitude*. Let S (Fig. 17) be the place of a star when on the meridian. Then its meridian altitude will be SH, which will best be found by taking its zenith distance ZS, of which it is the complement. From SH subtract EH, the elevation of the equator, which equals the co-latitude of the place of observation (Art. 44), and the remainder, SE, is the declination. Or, if the star is nearer the horizon than the equator is, as at S′, subtract its meridian altitude from the co-latitude, for the declination.

Fig. 17.

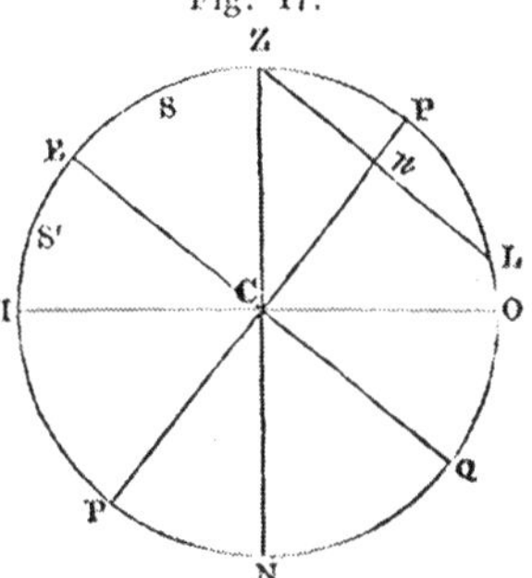

Secondly, the declination may be found from the *north polar distance*, of which it is the complement. Thus from P to E is 90°. Therefore, PE−PS=90°−PS=SE=the declination. The height of the pole P is always known when the latitude of the place is known, being equal to the latitude.

128. The astronomical instruments already described are adapted to taking observations on the meridian only, but we sometimes require to know the *altitude* of a celestial body when it is not on the meridian, and its *azimuth*, or distance from the meridian measured on the horizon; and also the *angular distance* between two points on any part of the celestial sphere. An instrument especially designed to measure altitudes and azimuths, is called an *Altitude and Azimuth Instrument*, whatever may be its particular form. When a point is on the horizon, its distance from the meridian, or its azimuth, may be taken by the common surveyor's compass, the direction of the meridian being determined by the needle; but when the object, as a star, is not on the horizon, its azimuth, it must be remembered, is the arc of the horizon from the meridian to a vertical circle passing through the star (Art. 27); at whatever different altitudes, therefore, two stars may be, and however the plane which passes through them may be inclined to the horizon, still it is their angular distance measured *on the horizon* which determines their difference of azimuth. Figure 18 represents an Altitude and Azimuth Instrument, several of the usual appendages and subordinate contrivances being omitted for the sake of distinctness and simplicity. Here *abc* is the vertical or altitude circle, and EFG the horizontal or azimuth circle; AB is a telescope mounted

Fig. 18.

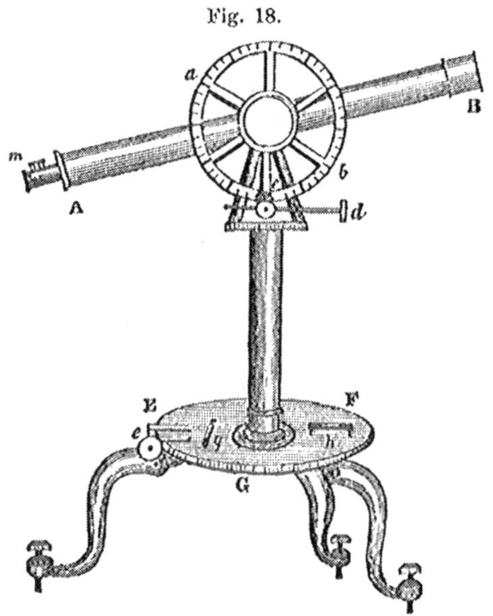

on a horizontal axis and capable of two motions, one in altitude parallel to the circle *abc*, and the other in azimuth parallel to EFG. Hence it can be easily brought to bear upon any object. At *m*, under the eye-glass of the telescope, is a small mirror placed at an angle of 45° with the axis of the telescope, by means of which the image of the object is reflected upward, so as to be conveniently presented to the eye of the observer. At *d* is represented a *tangent screw*, by which a slow motion is given to the telescope at *c*. At *h* and *g* are seen two *spirit-levels* at right angles to each other, which show when the azimuth circle is truly horizontal. The instrument is supported on a *tripod*, for the sake of greater steadiness, each foot being furnished with a screw for leveling.

129. The *sextant* is one of the most useful instruments, both to the astronomer and the navigator, and will therefore merit particular attention. In figure 19, I and II are two small mirrors, and T a small telescope. ID represents a movable arm, or radius, which carries an index at D. The radius turns on a pivot at I, and the index moves on a graduated arc EF.

Fig. 19.

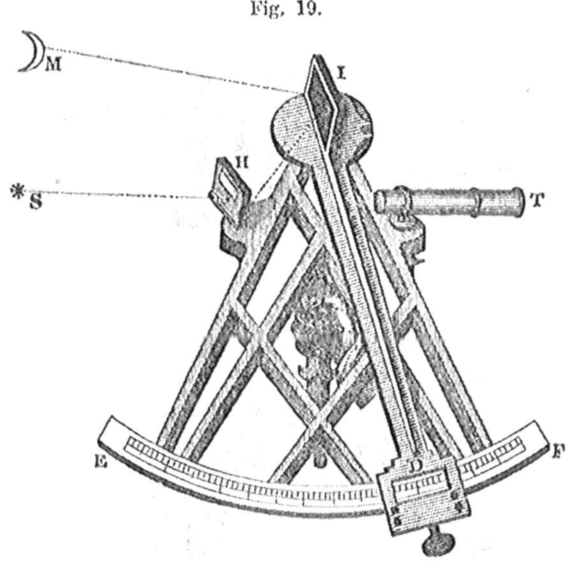

I is called the *Index Glass* and II the *Horizon Glass*. The under part only of the horizon glass is coated with quicksilver, the upper part being left transparent; so that while one object is seen through the upper part by direct vision, another may be seen through the lower part by reflection from the two mirrors. The instrument is so contrived, that when the index is moved up to F, where the zero-point is placed, or the graduation begins, the two reflectors I and II are exactly parallel to each other. If we now look through the telescope, T, so pointed as to see the star S through the transparent part of the horizon glass, we shall see the same star, in the same place, reflected from the silvered part; for the star (or any similar object) is at such a distance that the rays of light which strike upon the index glass I, are parallel to those which enter the eye directly, and will exhibit the object at the same place. Now, suppose we wish to measure the angular distance between two bodies, as the moon and a star, and let the star be at S and the moon at M. The telescope being still directed to S, turn the index arm ID from F toward E until the image of the moon is brought down to S, its lower limb just touching S. By a principle in optics, the angular distance between the moon and its image, is twice the angle between the mirrors. But the mirror has passed over the graduated arc FD; therefore double that arc is the angular distance between the star and the moon's *lower* limb. If we then bring the *upper* limb into contact with the star, the sum of both observations, divided by 2, will give the angular distance between the star and the moon's *center*. As each degree on the limb EF measures two degrees of angular distance, hence the divisions for single degrees are in fact only half a degree asunder; and a sextant, or the sixth part of the circle, measures an angular distance of 120°. The upper and lower points in the disk of the sun or of the moon, may be considered as two separate objects, whose distance from each other may be taken in a similar manner, and thus their apparent diameters at any time be determined. We may select our points of observation either in a vertical, or in a horizontal direction.

130. If we make a star, or the limb of the sun or moon, one of the objects, and the point in the horizon directly beneath, the other, we thus obtain the *altitude* of the object. In this observation, the horizon is viewed through the transparent part of the horizon-glass. At sea, where the horizon is usually well defined, the horizon itself may be used for taking altitudes; but on land, inequalities of the earth's surface oblige us to have recourse to an *artificial horizon.* This, in its simple state, is a basin of either water or quicksilver. By this means we see the image of the sun (or other body) just as far below the horizon as it is in reality above it. Hence, if we turn the index-glass until the limb of the sun, as reflected from it, is brought into contact with the image seen in the artificial horizon, we obtain double the altitude.*

* Woodhouse's Ast., p. 774.

The sextant must be held in such a manner that its plane shall pass through the plane of the two objects. It must be held, therefore, in a vertical plane in taking altitudes, and in a horizontal plane in taking the horizontal diameters of the sun and moon. Holding the instrument in the true plane of the two bodies, whose angular distance is measured, is indeed the most difficult part of the operation.

The peculiar value of the sextant consists in this, that the observations taken with it are not affected by any motion in the observer; hence it is the chief instrument used for angular measurements at sea.

131. *Examples illustrating the use of the Sextant.*

Ex. 1. Alt. ☉'s lower limb, . .	49° 10′ 00″
☉'s semi-diameter, . .	0 15 51
	49° 25′ 51″
Subtract Refraction, . .	00 00 49
	49° 25′ 02″
Add Parallax, . . .	00 00 06
True altitude ☉'s center, .	49° 25′ 08″

Ex. 2. *With the Artificial Horizon.*

Altitude of ☉'s upper limb above the image in the artificial horizon, 100° 2′ 47″.

True altitude, ...	50° 01′ 23″.5
☉'s semi-diameter, ...	00 15 50.
	49° 45′ 33″.5
Refraction, ...	00 00 48.
	49° 44′ 45″.5
Parallax, ...	00 00 05.
True altitude of ☉'s center, . .	49° 44′ 50″.5

ASTRONOMICAL PROBLEMS.*

132. *Given the sun's Right Ascension and Declination, to find his Longitude and the Obliquity of the Ecliptic.*

* Young's Spherical Trigonometry, p. 136. Vince's Complete System, vol. i.

Let PCP′ (Fig. 20) represent the celestial meridian that passes through the first of Cancer and Capricorn (the solstitial colure), PP′ the axis of the sphere, EQ the equator, E′C the ecliptic, and PSP′ the declination circle (Art. 37) passing through the sun S; then ARS is a right angle, and in the right-angled spherical triangle ARS, are given the right ascension AR (Art. 37), and the declination RS, to find the longitude AS and the obliquity SAR.

Fig. 20.

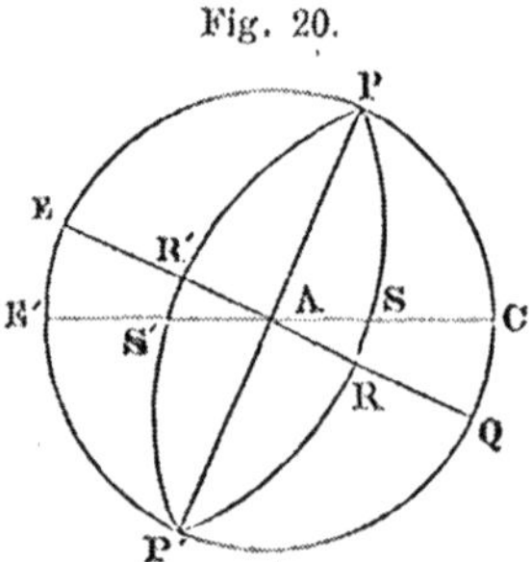

As longitude and right ascension are measured from A, the first point of Aries, in the direction AS of the signs, quite round the globe, when, of the four quantities mentioned in the problem, the obliquity and the declination are given to find the others, we must know whether the sun is north, or whether it is south of the equator, the longitude being in the one case AS, and in the other, instead of AS′, it is 360−AS′, when near the same equinox, A. We must also know on which side of the tropic the sun is, for the sun in passing from one of the tropics to the equinox, passes through the same degrees of declination as it had gone through in ascending from the other equinox to the tropic, although its longitude and right ascension go on continually increasing. From the 21st of March to the 21st of June, while describing the first quadrant from the vernal equinox, the declination is north and increasing; north, but decreasing, in the second quadrant, until the 23d of September; south and increasing in the third quadrant, until the 21st of December; and finally, in the fourth quadrant, south but decreasing until the 21st of March.

Ex. 1. On the 17th of May, the sun's Right Ascension was 53° 38′, and his Declination 19° 15′ 57″: required his Longitude and the Obliquity of the Ecliptic.

Applying *Napier's Rule** to the right-angled triangle ARS, we have

* The student is supposed to be acquainted with Spherical Trigonometry; but to refresh his memory, we may insert a remark or two.

1. Rad cos AS=cos AR cos RS.

2. Rad sin AR=tan RS cot A.·. cot A=$\frac{\text{rad sin AR}}{\text{tan RS}}$

Hence the computation for AS and A is as follows:

For the Longitude AS.		For the Obliquity A.	
cos AR 53° 38′ 00″	9.7730185	sin AR	9.9059247
cos RS 19 15 57	9.9749710	tan RS, ar. com.	0.4565209
cos AS 55 57 43	9.7479895	cot A 23° 27′ 50½″	10.3624456

It will be recollected that in Napier's rule for the solution of a right-angled spherical triangle, by means of the Five Circular Parts, we proceed as follows:

In a right-angled spherical triangle we are to recognize but five parts, viz., the three sides and the two oblique angles. If we take any one of these as a *middle part*, the two which lie next to it on each side will be *adjacent parts*. Thus (in Fig. 21), taking A for a middle part, *b* and *c* will be the adjacent parts; if we take *c* for the middle part, A and B will be the adjacent parts; if we take B for the middle part, *c* and *a* will be the adjacent parts; but if we take *a* for the middle part, then as the angle C is not considered as one of the circular parts, B and *b* are the adjacent parts; and, lastly, if *b* is the middle part, then the adjacent parts are A and *a*. The two parts immediately beyond the adjacent parts on each side, still disregarding the right angle, are called the *opposite parts*; thus, if A is the middle part, the opposite parts are *a* and B. Napier's rule is as follows:

Fig. 21.

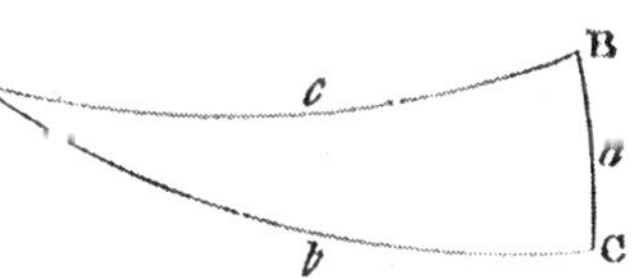

Radius into the sine of the middle part, equals the product of the tangents of the adjacent parts, or of the cosines of the opposite parts.

(The corresponding vowels are marked to aid the memory.) In the use of this rule, it must be understood that the *complements* of the angles and the hypotenuse are used, instead of those parts themselves. Thus, if A is middle part, we say rad × cos A, not rad × sin A, and so of B; or, if AB is adjacent part, we use cot AB, not tan AB; if opposite, sin AB, not cos AB, &c.

Examples. 1. In the right-angled triangle ABC, are given the two perpendicular sides, viz., a=48° 24′ 10″, b=59° 38′ 27″, to find the hypotenuse c. The hypotenuse being made the middle part, the other sides become the opposite parts, being separated from the middle part by the angles A and B. Hence,

$$\text{Rad}\cos c=\cos a\cos b\;.\because\;\cos c=\frac{\cos a\cos b}{\text{rad}}=70^\circ\ 23'\ 40''.$$

2. In the spherical triangle, right-angled at C, are given two perpendicular sides, viz., a=116° 30′ 43″, b=29° 41′ 32″, to find the angle A.

Here the required angle is *adjacent* to one of the given parts, viz., *b*, which make the middle part. Then,

$$\text{Rad}\times\sin b=\cot A\tan a\;.\because\;\cot A=\frac{\text{rad}\times\sin b}{\tan a}=76^\circ\ 7'\ 13''.$$

Ex. 2. On the 31st of March, 1816, the sun's Declination was observed at Greenwich to be 4° 13′ 31½″: required his Right Ascension, the obliquity of the ecliptic being 23° 27′ 51″.

Ans. 9° 47′ 59″.

Ex. 3. What was the sun's Longitude on the 28th of November, 1810, when his Declination was 21° 16′ 4″, and his Right Ascension, in time, 16h. 14m. 58.4s.?

Ans. 245° 39′ 10″.

Ex. 4. The sun's Longitude being 8ˢ 7° 40′ 56″, and the Obliquity 23° 27′ 42¼″, what was the Right Ascension in time? *Ans.* 16h. 23m. 34s.

133. *Given the sun's Declination to find the time of his Rising and Setting at any place whose latitude is known.*

Let PEP′ (Fig. 22) represent the meridian of the place, Z being the zenith, and HO the horizon; and let LL′ be the apparent path of the sun on the proposed day, cutting the horizon in S. Then the arc EZ will be the latitude of the place, and consequently EH, or its equal QO, will be the co-latitude, and this measures the angle OAQ; also RS will be the sun's declination, and AR expressed in time will be the time of rising before 6 o'clock. For it is evident that it will be sunrise when the sun arrives at the horizon at S; but PP′ being an hour circle whose plane is perpendicular to the meridian (and of course projected into a straight line on the plane of projection), the time the sun is passing from S to S′ taken from the time of describing S′L, which is six hours, must be the time from midnight to sunrise. But the time of describing SS′ is measured on the corresponding arc of the equinoctial AR.

Fig. 22.

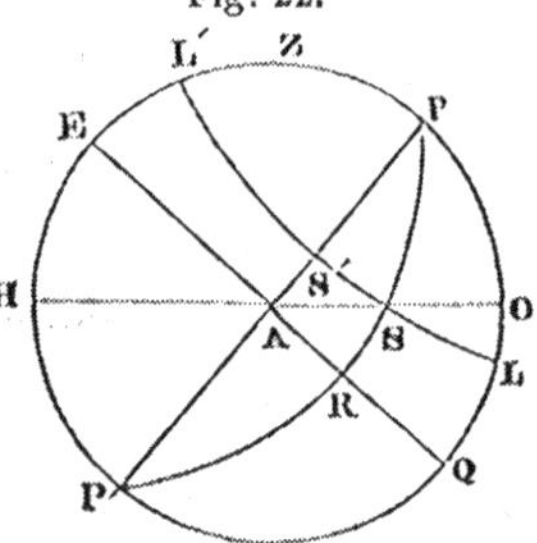

In the right-angled triangle ARS, we have the declination RS, and the angle A to find AR. Therefore,

$$\text{Rad} \times \sin \text{AR} = \cot \text{A} \times \tan \text{RS}.$$

Ex. 1. Required the time of sunrise at latitude 52° 13′ N. when the sun's declination is 23° 28′.

Rad		10.
Cot A or tan	52° 13′	10.1105786
Tan RS=	23° 28′	9.6376106
Sin	34° 03′ 21¼″	9.7481892
	4*	
	2h. 16′ 13″ 25‴	
	6	

3h. 43′ 46″ 35‴=the time after midnight, and of course the time of rising.

Ex. 2. Required the time of sunrise at latitude 57° 2′ 54″ N. when the sun's declination is 23° 28′ N.

Ans. 3h. 11m. 49s.

Ex. 3. How long is the sun above the horizon in latitude 58° 12′ N. when his declination is 18° 40′ S.?

Ans. 7h. 35m. 52s.

134. *Given the Latitude of the place, and the Declination of a heavenly body, to determine its Altitude and Azimuth when on the six o'clock hour circle.*

Let HZO (Fig. 23) be the meridian of the place, Z the zenith, HO the horizon, S the place of the object on the 6 o'clock hour circle PSP′, which of course cuts the equator in the east and west points, and ZSB the vertical circle passing through the body. Then in the right-angled triangle SBA, the given quantities are AS, which is the declination, and the arc OP or angle SAB, the latitude of the place, to find the altitude BS, and the azimuth BO, or the amplitude AB, which is its complement.

Fig. 23.

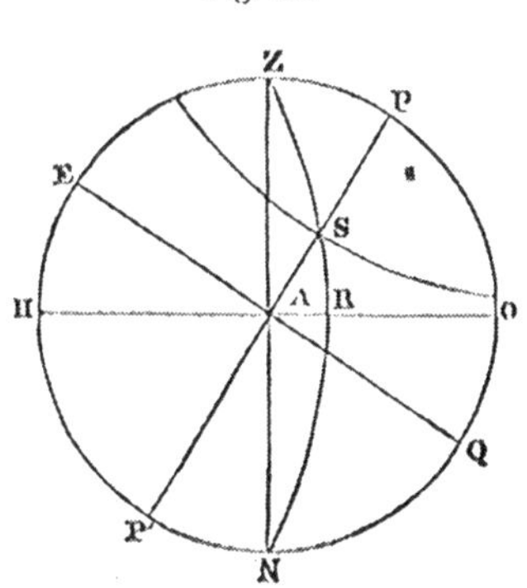

Ex. 1. What were the altitude and azimuth of Arcturus, when upon the six o'clock hour circle of Greenwich, lat. 51°

* Degrees are converted into hours by multiplying by 4 and dividing by 60.

28′ 40″ N. on the first of April, 1822; its declination being 20° 6′ 50″ N.?

For the Altitude.		For the Azimuth.	
Rad sin BS=sin AS sin A		Rad cos A=cot BO cot AS	
Rad . .	10.	Cot 20° 06′ 50″	10.4362545
Sin 20° 06′ 50″	9.5364162	Cos 51 28 40	9.7943612
Sin 51 28 40	9.8934103	Rad . .	10.
Sin 15 36 27	9.4298265	Cot 77° 09′ 04″	9.3581067

Ex. 2. At latitude 62° 12′ N. the altitude of the sun at 6 o'clock in the morning was found to be 18° 20′ 23″: required his declination and azimuth.

Ans. Dec. 20° 50′ 12″ N. Az. 79° 56′ 4″.

135. *The Latitudes and Longitudes of two celestial objects being given, to find their Distance apart.*

Let P (Fig. 24) represent the pole of the ecliptic, and PS, PS′, two arcs of celestial latitude (Art. 37) drawn to the two objects SS′; then will these arcs represent the co-latitudes, the angle P will be the difference of longitude, and the arc SS′ will be the distance sought. Here we have the two sides and the included angle given to find the third side. By Napier's Rules for the solution of oblique-angled spherical triangles (see Spherical Trigonometry), the sum and difference of the two angles opposite the given sides may be found, and thence the angles themselves. The required side may then be found by the theorem, that the sines of the sides are as the sines of their opposite angles.* The computation is omitted here on account of its great length. If P be the pole of the *equator* instead of the ecliptic, then PS and PS′ will represent arcs of co-declination, and the angle P will denote difference of right ascension. From these data, also, we may therefore derive the distance between any two stars. Or, finally, if P be the pole of the *horizon*, the angle at P will denote difference of azimuth, and

Fig. 24.

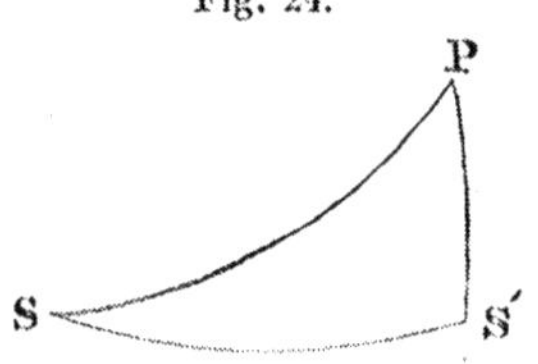

* More concise formulæ for the solution of this case may be found in Young's Trigonometry, p. 99; Francœur's Uranography, Art. 330; Dr. Bowditch's Practical Navigator, p. 436.

the sides PS, PS′, zenith distances, from which the side SS′ may likewise be determined.

FIGURE AND DENSITY OF THE EARTH.

136. We have already shown (Art. 8) that the figure of the earth is *nearly* globular; but since the semi-diameter of the earth is taken as the base line in determining the parallax of the heavenly bodies, and lies, therefore, at the foundation of all astronomical measurements, it is very important that it should be ascertained with the greatest possible exactness. Having now learned the use of astronomical instruments, and the method of measuring arcs on the celestial sphere, we are prepared to understand the methods employed to determine the exact figure of the earth. This element is indeed ascertained in five different ways, each of which is independent of all the rest, namely, by investigating the effects of the *centrifugal force* arising from the revolution of the earth on its axis—by measuring *arcs of the meridian*—by experiments with the *pendulum*—by the unequal action of the earth on the *moon*, arising from the redundance of matter about the equatorial regions—and by the *precession of the equinoxes.* We will briefly consider each of these methods.

137. First, *the known effects of the centrifugal force would give to the earth a spheroidal figure, elevated in the equatorial, and flattened in the polar regions.*

Had the earth been originally constituted (as geologists suppose) of yielding materials, either fluid or semi-fluid, so that its particles could obey their mutual attraction, while the body remained at rest it would spontaneously assume the figure of a perfect sphere; as soon, however, as it began to revolve on its axis, the greater velocity of the equatorial regions would give to them a greater centrifugal force, and cause the body to swell out into the form of an oblate spheroid.* Even had the solid part of the earth consisted of unyielding materials and been created a perfect sphere, still the waters that covered it would have receded from the polar and have been accumulated

* See a good explanation of this subject in the Edinburgh Encyclopædia, ii., 665.

in the equatorial regions, leaving bare extensive regions on the one side, and ascending to a mountainous elevation on the other.

On estimating from the known dimensions of the earth and the velocity of its rotation, the amount of the centrifugal force in different latitudes, and the figure of equilibrium which would result, Newton inferred that the earth must have the form of an oblate spheroid before the fact had been established by observation; and he assigned nearly the true ratio of the polar and equatorial diameters.

138. Secondly, *the spheroidal figure of the earth is proved by actually measuring the length of a degree on the meridian in different latitudes.*

Were the earth a perfect sphere, the section of it made by a plane passing through its center in any direction would be a perfect circle, whose curvature would be equal in all parts; but if we find by actual observation that the curvature of the section is not uniform, we infer a corresponding departure in the earth from the figure of a perfect sphere. This task of measuring portions of the meridian has been executed in different countries by means of a system of triangles with astonishing accuracy.* The result is, that the length of a degree increases as we proceed from the equator toward the pole, as may be seen from the following table:

Places of observation.	Latitude.	Length of a degree in miles.
Peru	00° 00′ 00″	68.732
Pennsylvania ...	39 12 00	68.896
Italy	43 01 00	68.998
France.........	46 12 00	69.054
England	51 29 54½	69.146
Sweden	66 20 10	69.292

Combining the results of various measurements, the dimensions of the terrestrial spheroid are found to be as follows:†

Equatorial diameter, ... 7925.308
Polar diameter, ... 7898.952
Mean diameter, ... 7912.130

* See Day's Trigonometry.

† Bessel.

The difference between the greatest and least, is $26.356 = \frac{1}{301}$ of the greatest. This fraction ($\frac{1}{301}$) is denominated the *ellipticity* of the earth, being the excess of the transverse over the conjugate axis, on the supposition that the section of the earth coinciding with the meridian is an ellipse; and that such is the case, is proved by the fact that calculations on this hypothesis, of the lengths of arcs of the meridian in different latitudes, agree nearly with the lengths obtained by actual measurement.

139. Thirdly, *the figure of the earth is shown to be spheroidal by observations with the pendulum.*

The use of the pendulum in determining the figure of the earth, is founded upon the principle that *the number of vibrations performed by the same pendulum, when acted on by different forces, varies as the square root of the forces.** Hence, by carrying a pendulum to different parts of the earth, and counting the number of vibrations it performs in a given time, we obtain the relative forces of gravity at those places; and this leads to a knowledge of the relative distance of each place from the center of the earth, and finally, to the ratio between the equatorial and the polar diameters.

140. Fourthly, *that the earth is of a spheroidal figure is inferred from the motions of the moon.*

These are found to be affected by the excess of matter about the equatorial regions, producing certain irregularities in the lunar motions, the amount of which becomes a measure of the excess itself, and hence affords the means of determining the earth's ellipticity. This calculation has been made by the most profound mathematicians, and the figure deduced from this source corresponds very nearly to that derived from the several other independent methods.

141. Fifthly, *the spheroidal form of the earth accounts for the precession of the equinoxes.*

It will be shown in Chap. IV., Part II., that the slow backward motion of the equinoctial points on the ecliptic is due to the attraction of the sun and moon upon the belt of matter on

* Mechanics, Art. 161.

the equator, combined with the inertia of the earth in its rotation on its axis.

We thus have the shape of the earth established upon the most satisfactory evidence, and are furnished with a starting point from which to determine various measurements among the heavenly bodies.

141′. The *density* of the earth compared with water, that is, its specific gravity, is $5\frac{1}{2}$.* The density was first estimated by Dr. Hutton, from observations made by Dr. Maskelyne, Astronomer Royal, on Schehallien, a mountain of Scotland, in the year 1774. Thus, let M (Fig. 25) represent the mountain, D, B, two stations on opposite sides of the mountain, and I a star; and let IE and IG be the zenith distances as determined by the differences of latitudes of the two stations. But the apparent zenith distances as determined by the plumb-line are IE′ and IG′. The deviation toward the mountain on each side exceeded 7″.† The attraction of the mountain being observed on both sides of it, and its mass being computed from the number of sections taken in all directions, these data, when compared with the known attraction and magnitude of the earth, led to a knowledge of its mean density. According to Dr. Hutton, this is to that of water as 9 to 2; but later and more accurate estimates have made the specific gravity of the earth as stated above. But this density is nearly double the average density of the materials that compose the exterior crust of the earth, showing a great increase of density toward the center.

Fig. 25.

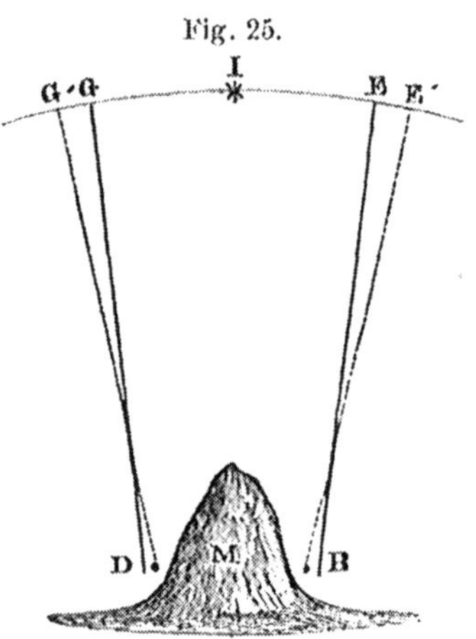

The density of the earth is an important element, as we shall find that it helps us to a knowledge of the density of each of the other members of the solar system.

---

* Baily, Ast. Tables, p. 21.

† Robison's Phys. Ast.

PART II.—OF THE SOLAR SYSTEM.

142. Having considered the Earth, in its astronomical relations, and the Doctrine of the Sphere, we proceed now to a survey of the Solar System, and shall treat successively of the Sun, Moon, Planets, and Comets.

CHAPTER I.

OF THE SUN—SOLAR SPOTS—ZODIACAL LIGHT.

143. The *figure* which the sun presents to us is that of a perfect circle, whereas most of the planets exhibit a disk more or less elliptical, indicating that the true shape of the body is an oblate spheroid. So great, however, is the distance of the sun, that a line 400 miles long would subtend an angle of only 1″ at the eye, and would, therefore, be the least space that could be measured. Hence, were the difference between two conjugate diameters of the sun any quantity less than this, we could not determine by actual measurement that it existed at all. Still we learn from theoretical considerations, founded upon the known effects of centrifugal force, arising from the sun's revolution on his axis, that his figure is not a perfect sphere, but is slightly spheroidal.*

144. The *distance of the sun from the earth* is nearly 95,000,000 miles. For, its horizontal parallax being 8″.6 (Art. 86), and the semi-diameter of the earth 3956 miles,

Sin 8″.6 : 3956 : : Rad : 95,000,000 nearly.

In order to form some faint conception at least of this vast distance, let us reflect that a railway car, moving at the rate of 20 miles per hour, would require more than 500 years to reach the sun.

* See Mécanique Céleste, iii., p. 165. Delambre, t. i., p. 483.

145. The apparent *diameter* of the sun may be found either by the Sextant (Art. 129), by an instrument called the *Heliometer*, specially designed for measuring its angular breadth, or by the time it occupies in crossing the meridian.

The last is the most accurate. If the sun, when on the equator, March 21 or September 22, is found to cross the meridian in 2m. 10s., sidereal time, then 24h. : 2m. 10s. : : 360° : 32′ 30″ = the angular breadth of the sun. But if the sun has a declination north or south, the degrees of the diurnal circle on which he moves are shorter than those on the equator, in the ratio of the cosine of declination to radius; and therefore the time of crossing is lengthened; hence the calculated breadth must be diminished in the same ratio. Having the distance and angular diameter, we can easily find its *linear* diameter. Let E (Fig. 26) be the earth, S the sun, ES a line drawn to the center of the disk, and EC a line drawn touching the disk at C. Join SC; then

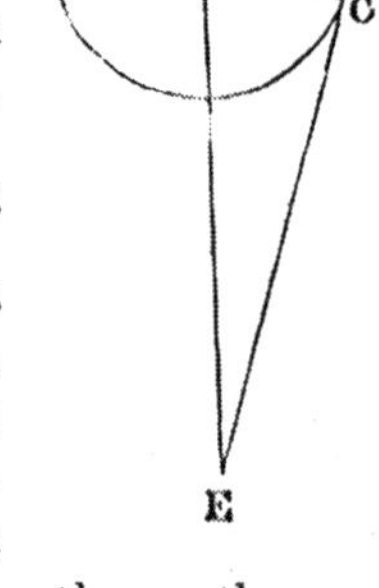

Fig. 26.

Rad : ES : : sin 16′ 1″.5 (the sun's mean apparent semidiameter) : SC = 442,840 miles.

And $\frac{2 \times 442840}{7912} = 112$ nearly; that is, it would require one hundred and twelve bodies like the earth, if laid side by side, to reach across the diameter of the sun. Since spheres are to each other as the cubes of their diameters, $1^3 : 112^3$: : 1 : 1,400,000 nearly; that is, the sun is about 1,400,000 times as large as the earth.

146. In *density*, the sun is only one-fourth that of the earth, being but a little heavier than water (Art. 141′); and since the quantity of matter, or *mass* of a body, is proportioned to its magnitude and density, hence $1,400,000 \times \frac{1}{4} = 350,000$; that is, the quantity of matter in the sun is three hundred and fifty thousand (or, more accurately, 354,936) times as great as in the earth. Now the weight of bodies (which is a measure of the force of gravity) varies directly as the quantity of matter, and inversely as the square of the distance. A body, therefore, would weigh 350,000 times as much on the surface of the sun as on the earth, if the distance of the center of force were the same in both cases; but since the attraction of a sphere is the same as though all the matter were collected in the center, consequently, the weight of a body, so far as it depends on its distance from the center of force, would be the square of 112 times less at the sun than at the earth. Or, putting W for the weight at the earth, and W′ for the weight at the sun, then

$$W : W' :: \frac{1}{1^2} : \frac{350,000}{(112)^2} = 27.9 \text{ lbs.}$$

Hence a body would weigh nearly 28 times as much at the sun as at the earth. A man weighing 200 lbs. would, if transported to the surface of the sun, weigh 5,580 lbs., or nearly $2\frac{1}{2}$ tons. To lift one's limbs would, in such a case, be beyond the ordinary power of the muscles. At the surface of the earth a body falls through $16\frac{1}{12}$ feet in a second; and since the spaces are as the velocities, the times being equal, and the velocities as the forces, therefore a body would fall at the sun in one second through $16\frac{1}{12} \times 27\frac{9}{10} = 448.7$ feet.

SOLAR SPOTS.

147. The surface of the sun, when viewed with a telescope, often shows dark spots, which vary much, at different times, in number, figure, and extent. One hundred or more, assembled in several distinct groups, are sometimes visible at once on the solar disk. The solar spots are commonly very small, but occasionally a spot of enormous size is seen occupying an extent of 50,000 miles or more in diameter. They are sometimes even visible to the naked eye, when the sun is viewed through colored glass, or when near the horizon, it is seen through light clouds or vapors. When it is recollected that 1″ of the solar disk implies an extent of 400 miles (Art. 143), it is evident that a space large enough to be seen by the naked eye, must cover a very large extent.

Fig. 27 exhibits the appearance of solar spots, though much too large compared with the disk. Whenever a spot is seen near the edge of the disk, it appears foreshortened by perspective, as in the figure. A solar spot usually consists of two parts, the *nucleus* and the *umbra*. The nucleus is black, of a very irregular shape, and is subject to great and sudden changes

Fig. 27.

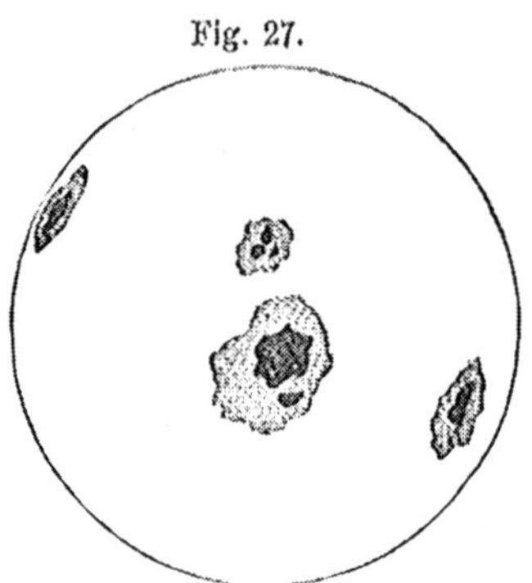

both in form and size. A spot sometimes divides into many smaller ones, and again a group may be united into a single spot. The umbra is a wide margin of lighter shade, and is commonly of greater extent than the nucleus. The spots are usually confined to a zone extending across the central regions of the sun, not exceeding 60° in breadth. When the spots are observed from day to day, they are seen *to move across the disk of the sun*, occupying about two weeks in passing from one limb to the other. After an absence of about the same period, the spot returns, having taken 27d. 7h. 37m. in the entire revolution.

148. Besides the fact of foreshortening already mentioned, there is another proof that the spots are *at the surface* of the sun. Were they bodies at a distance from it, the time during which they would be seen on the solar disk would be less than that occupied in the remainder of the revolution. Thus, let S (Fig. 28) be the sun, E the earth, and *abc* the path of the body revolving about the sun. Unless the spot were nearly or quite in contact with the body of the sun, being projected upon his disk only while passing from *b* to *c*, and being invisible while describing the arc *cab*, it would of course be out of sight longer than in sight, whereas the two periods are found to be equal. Moreover, the lines which all the solar spots describe on the disk of the sun are found to be parallel to each other, like the circles of diurnal revolution around the earth; and hence it is inferred that they arise from a similar cause, namely, *the revolu-*

Fig. 28.

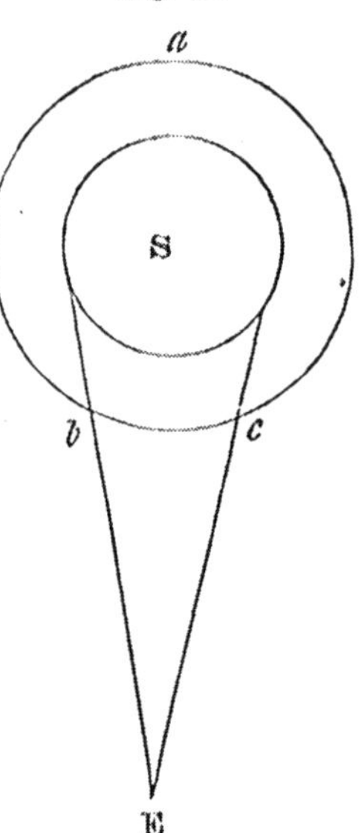

tion of the sun on his axis, a fact which is thus made known to us.

But although the spots occupy about 27¼ days in passing from one limb of the sun around to the same limb again, yet this is not the period of the sun's revolution on his axis, but exceeds it by nearly two days. For, let AA′B (Fig. 29) represent the sun, and EE′M the orbit of the earth. When the earth is at E, the visible disk of the sun will be AA′B; and if the earth remained stationary at E, the time occupied by a spot after leaving A until it returned to A, would be just equal to the time of the sun's revolution on his axis. But during the 27¼ days in which the spot has been performing its apparent revolution, the earth has been advancing in her orbit from E to E′, where the visible disk of the sun is A′B′. Consequently, before the spot can appear again on the limb from which it set out, it must describe so much more than an entire revolution as equals the arc AA′, which equals the arc EE′. Hence,

Fig. 29.

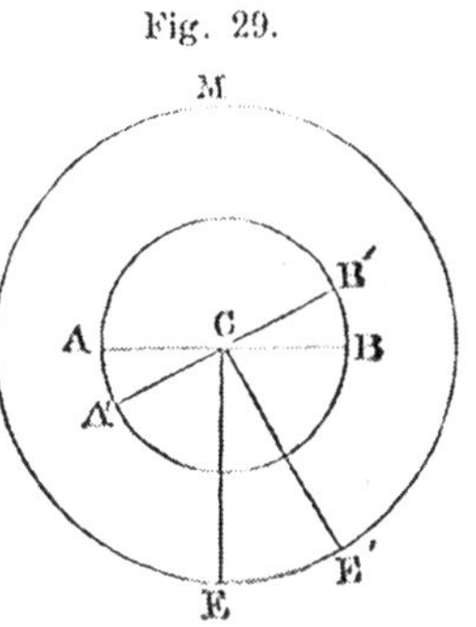

365d. 5h. 48m.+27d. 7h. 37m. : 365d. 5h. 48m. :: 27d. 7h. 37m. : 25d. 9h. 59m.=the time of the sun's revolution on his axis.

149. If the path which the spots appear to describe by the revolution of the sun on his axis left each a visible trace on his surface, they would form, like the circles of diurnal revolution on the earth, so many parallel rings, of which that which passed through the center would constitute the solar equator, while those on each side of this great circle would be small circles, corresponding to parallels of latitude on the earth. Let us conceive of an artificial sphere to represent the sun, having such rings plainly marked on its surface. Let this sphere be placed at some distance from the eye, with its axis perpendicular to the axis of vision, in which case the equator would coincide with the line of vision, and its edge be presented to the eye. It would therefore be projected into a straight line. The same would be the case with all the smaller rings, the distance being supposed such that the rays of light come from them all to the eye nearly parallel. Now let the axis, instead of being perpendicular to the line of vision, be inclined to that line, then all the rings being seen obliquely, would be projected into ellipses. If, however, while the sphere remained in a fixed position, the eye were carried around it (being always in the same plane) twice during the circuit, it would be in the plane of the equator, and project this and all the smaller circles into straight lines; and twice, at points 90° distant from the foregoing positions, the eye would be at a distance from the planes of the rings equal to the inclination of the equator of the sphere to the line of vision. Here it would project the rings into wider ellipses than at other points; and the ellipses would become more and more eccentric as the eye departed from either of these points, until they vanished again into straight lines.

150. It is in a similar manner that the eye views the paths described by the spots on the sun. If the sun revolved on an axis perpendicular to the plane of the earth's orbit, the eye being situated in the plane of revolution, and at such a distance from the sun that the light comes to the eye from all parts of the solar disk nearly parallel, the paths described by the spots would be projected into straight lines, and each would describe a straight line across the solar disk, parallel to the plane of revolution. But the axis of the sun is inclined to the ecliptic about 7¼° from a perpendicular, so that usually all the circles described by the spots are projected into ellipses. The breadth of these, however, will vary as the eye, in the annual revolution, is carried around the sun, and when the eye comes into the plane of the rings, as it does twice a year, they are projected into straight lines, and for a short time a spot seems moving in a straight line inclined to the plane of the ecliptic 7¼°. The two points where the sun's equator cuts the ecliptic are called the sun's *nodes*. The longitudes of the nodes are 80° 7′ and 260° 7′, and the earth passes through them about the 12th of December, and the 11th of June. It is at these times that the spots appear to describe straight lines. We have mentioned the various changes in the apparent paths of the solar spots, which arise from the inclination of the sun's axis to the plane of the ecliptic; but it was in fact by first observing these changes, and proceeding in the reverse order from that which we have pursued, that astronomers ascertained that the sun revolves on his axis, and that this axis is inclined to the ecliptic 82¾°.

151. Besides the spots already described, there are faint inequalities of light over the general surface, in delicate lines and freckles, which are also perpetually changing. These are called *faculæ*.

The theory, which is generally received, regards the visible surface of the sun as an incandescent, gaseous substance, always in violent commotion, and sometimes rent here and there by a broad opening, which reveals a lower stratum of less illumination. This is the *umbra* of a spot; and a smaller rupture in *that* shows a still lower stratum, or else the solid body of the sun, as the *nucleus* of the same. These apertures in the luminous strata may be caused, as some think, by volcanic action below, or according to others, by storms or tornadoes in the solar atmosphere.

ZODIACAL LIGHT.

152. The Zodiacal Light is a faint light resembling the tail of a comet, and is seen at certain seasons of the year following the course of the sun after evening twilight, or preceding his approach in the morning sky. Figure 30 represents its appearance as seen in the evening, in March, 1836. The following are the leading facts respecting it.

Fig. 30.

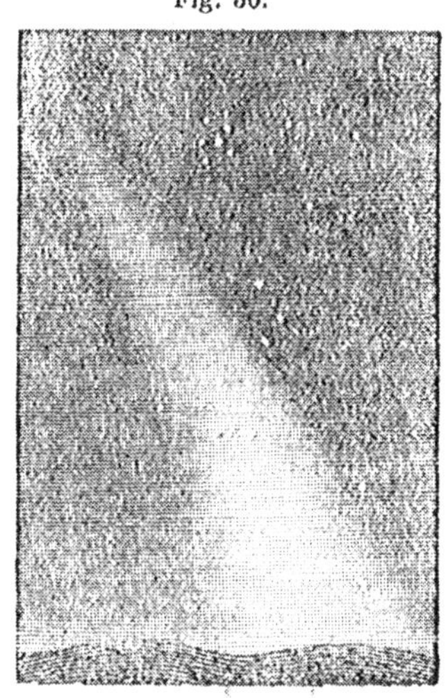

1. *Its form is that of a luminous triangle*, having its base toward the sun. It reaches to an immense distance from the sun, sometimes even beyond the

orbit of the earth. It is brighter in the parts nearer the sun than in those that are more remote, and terminates in an obtuse apex, its light fading away by insensible gradations, until it becomes too feeble for distinct vision. Hence its limits are, at the same time, fixed at different distances from the sun by different observers, according to their respective powers of vision.

2. *Its aspects vary very much with the different seasons of the year.* About the first of October, in our climate (Lat. 41° 18′), it becomes visible before the dawn of day, rising along north of the ecliptic, and terminating above the nebula of Cancer. About the middle of November, its vertex is in the constellation Leo. At this time no traces of it are seen in the west after sunset, but about the first of December it becomes faintly visible in the west, crossing the Milky Way near the horizon, and reaching from the sun to the head of Capricornus, forming, as its brightness increases, a counterpart to the Milky Way, between which on the right, and the Zodiacal Light on the left, lies a triangular space embracing the Dolphin. Through the month of December, the Zodiacal Light is seen on both sides of the sun, namely, before the morning and after the evening twilight, sometimes extending 50° westward, and 70° eastward of the sun at the same time. After it begins to appear in the western sky, it increases rapidly from night to night, both in length and brightness, and withdraws itself from the morning sky, where it is scarcely seen after the month of December, until the next October.

3. *The Zodiacal Light moves through the heavens in the order of the signs.* It moves with unequal velocity, being sometimes stationary and sometimes retrograde, while at other times it advances much faster than the sun. In February and March, it is very conspicuous in the west, reaching to the Pleiades and beyond; but in April it becomes more faint, and nearly or quite disappears during the month of May. It is scarcely seen in this latitude during the summer months.

4. *It is remarkably conspicuous at certain periods of a few years, and then for a long interval almost disappears.*

5. *The Zodiacal Light was formerly held to be the atmosphere of the sun.** But La Place has shown that the solar

* Mairan, Memoirs French Academy, for 1733.

atmosphere could never reach so far from the sun as this light is seen to extend.* It has been supposed by others to be a nebulous body revolving around the sun. From recent observations, made with care in various parts of tropical America, there appears to be strong evidence that the Zodiacal Light is a belt which entirely surrounds the earth.†

CHAPTER II.

OF THE APPARENT ANNUAL MOTION OF THE SUN—SEASONS—FIGURE OF THE EARTH'S ORBIT.

153. THE revolution of the earth around the sun once a year, produces an apparent motion of the sun around the earth in the same period. When bodies are at such a distance from each other as the earth and the sun, a spectator on either would project the other body upon the concave sphere of the heavens, always seeing it on the opposite side of a great circle, 180° from himself. Thus, when the earth arrives at Libra (Fig. 11), we see the sun in the opposite sign Aries. When the earth moves from Libra to Scorpio, as we are unconscious of our own motion, the sun it is that appears to move from Aries to Taurus, being always seen in the heavens, where a line drawn from the eye of the spectator through the body meets the concave sphere of the heavens. Hence the line of projection carries the sun forward on one side of the ecliptic, at the same rate as the earth moves on the opposite side; and therefore, although we are unconscious of our own motion, we can read it from day to day in the motions of the sun. If we could see the stars at the same time with the sun, we could actually observe from day to day the sun's progress through them, as we observe the progress of the moon at night; only the sun's rate of motion would be nearly fourteen times slower than that of the moon. Although we do not see the stars

* Mec. Céleste. iii., 525.

† See a paper by Rev. Geo. Jones, U. S. Navy. Proc. Amer. Assoc., 1859, p. 172.

when the sun is present, yet after the sun is set, we can observe that it makes daily progress eastward, as is apparent from the constellations of the Zodiac occupying, successively, the western sky after sunset, proving that either all the stars have a common motion westward independent of their diurnal motion, or that the sun has a motion past them, from west to east. We shall see hereafter abundant evidence to prove that this change in the relative position of the sun and stars, is owing to a parallactic change in the place of the sun, and not to any change in the stars.

154. Although the apparent revolution of the sun is in a direction opposite to the real motion of the earth, as regards absolute space, yet both are nevertheless from west to east, since these terms do not refer to any directions in absolute space, but to the order in which certain constellations (the constellations of the Zodiac) succeed one another. The earth itself, on opposite sides of its orbit, does, in fact, move toward directly opposite points of space; but it is all the while pursuing its course in the order of the signs. In the same manner, although the earth turns on its axis from west to east, yet any place on the surface of the earth is moving in a direction in space exactly opposite to its direction twelve hours before. If the sun left a visible trace on the face of the sky, the ecliptic would, of course, be distinctly marked on the celestial sphere as it is on an artificial globe; and were the equator delineated in a similar manner (by any method like that supposed in Art. 46), we should then see at a glance the relative position of these two circles; the points where they intersect one another constituting the equinoxes, the points where they are at the greatest distance asunder, or the solstices, and various other particulars, which, for want of such visible traces, we are now obliged to search for by indirect and circuitous methods. It will even aid the learner to have constantly before his mental vision, an imaginary delineation of these two important circles on the face of the sky.

155. *The method of ascertaining the nature and position of the earth's orbit is by observations on the sun's Declination and Right Ascension.*

The exact declination of the sun at any time is determined from his *meridian altitude* or *zenith distance*, the latitude of the place of observation being known (Art. 37). The instant the center of the sun is on the meridian (which instant is given by the transit instrument), we take the distance of his upper and that of his lower limb from the zenith: half the sum of the two observations corrected for refraction, gives the zenith distance of the center. This result is diminished for parallax (Art. 84), and we obtain the zenith distance as it would be if seen from the center of the earth. The zenith distance being known, the declination is readily found by subtracting that distance from the latitude. By thus taking the sun's declination for every day of the year at noon, and comparing the results, we learn its motion *to and from the equator*.

156. To obtain the motion in right ascension, we observe, with a transit instrument, the instant when the center of the sun is on the meridian. Our sidereal clock gives us the right ascension in time (Art. 124), which we may easily, if we choose, convert into degrees and minutes, although it is more common to express right ascension by hours, minutes, and seconds. The differences of right ascension from day to day throughout the year, give us the sun's annual motion *parallel to the equator*. From the daily records of these two motions, at right angles to each other, arranged in a table,* it is easy to trace out the path of the sun on the artificial globe; or to calculate it with the greatest precision by means of spherical triangles, since the declination and right ascension constitute two sides of a right-angled spherical triangle, the corresponding arc of the ecliptic, that is, the longitude, being the third side (Art. 132). By inspecting a table of observations, we shall find that the declination attains its greatest value on the 22d of December, when it is 23° 27′ 54″ south; that from this period it diminishes daily and becomes nothing on the 21st of March; that it then increases toward the north, and reaches a similar maximum at the northern tropic about the 22d of June; and, finally, that it returns again to the southern tropic by grada-

* Such a table may be found in Biot's Astronomy, in Delambre, and in most collections of Astronomical Tables.

tions similar to those which marked its northward progress. A table of observations, also, would show us that the daily differences of declination are very unequal; that, for several days, when the sun is near either tropic, its declination scarcely varies at all; while near the equator, the variations from day to day are very rapid—a fact which is easily understood, when we reflect, that at the solstices the equator and the ecliptic are parallel to each other,* both being at right angles to the meridian; while at the equinoxes, the ecliptic departs most rapidly from the direction of the equator.

On examining, in like manner, a table of observations of the right ascension, we find that the daily differences of right ascension are likewise unequal; that the mean of them all is $3^m\ 56^s$, or 236^s, but that they have varied between 215^s and 266^s. On examining, moreover, the right ascension at each of the equinoxes, we find that the two records differ by 180°; which proves that the path of the sun is a great circle, since no other would bisect the equinoctial as this does.

157. *The obliquity of the ecliptic is equal to the sun's greatest declination.* For, by article 22, the inclination of any two great circles is equal to their greatest distance asunder, as measured on the sphere. The obliquity of the ecliptic may be determined from the sun's meridian altitude, or zenith distance, on the day of the solstice. The exact instant of the solstice, however, is not likely to occur when the sun is on the meridian, but may happen at some other meridian; still, the changes of declination near the solstice are so exceedingly small that but a slight error can result from this source. The obliquity may also be found, without knowing the latitude, by observing the greatest and least meridian altitudes of the sun, and taking half the difference. This is the method practiced in ancient times by Hipparchus. (Art. 2.) On comparing observations made at different periods for more than two thousand years, it is found that the obliquity of the ecliptic is not constant, but that it undergoes a slight diminution from age to age, amounting to 52″ in a century, or about half a second annually. We

* Or, more properly, the *tangents* of the two circles (which denote the directions of the curves at those points) are parallel.

might apprehend that by successive approaches to each other the equator and ecliptic would finally coincide; but astronomers have ascertained by an investigation, founded on the principles of universal gravitation, that this variation is confined within certain narrow limits, and that the obliquity, after diminishing for some thousands of years, will then increase for a similar period, and will thus vibrate forever about a mean value.

158. *The dimensions of the earth's orbit, when compared with its own magnitude are immense.*

Since the distance of the earth from the sun is 95,000,000 miles, and the length of the entire orbit nearly 600,000,000 miles, it will be found, on calculation, that the earth moves 1,640,000 miles per day, 68,000 miles per hour, 1,100 miles per minute, and nearly 19 miles every second, a velocity nearly fifty times as great as the maximum velocity of a cannon-ball. A place on the earth's equator turns, in the diurnal revolution, at the rate of about 1,000 miles an hour, and $\frac{5}{18}$ of a mile per second. The motion around the sun, therefore, is nearly 70 times as swift as the greatest motion around the axis.

THE SEASONS.

159. *The change of seasons depends on two causes*, (1) *the obliquity of the ecliptic, and* (2) *the earth's axis always remaining parallel to itself.* Had the earth's axis been perpendicular to the plane of its orbit, the equator would have coincided with the ecliptic, and the sun would have constantly appeared in the equator. To the inhabitants of the equatorial regions, the sun would always have appeared to move in the prime vertical; and to the inhabitants of either pole, he would always have been in the horizon. But the axis being turned out of a perpendicular direction 23° 28′, the equator is turned the same distance out of the ecliptic; and since the equator and ecliptic are two great circles which cut each other in two opposite points, the sun, while performing his circuit in the ecliptic, must, evidently, be once a year in each of those points, and must depart from the equator of the heavens to a distance on either side equal to the inclination of the two circles; that is, 23° 28′. (Art. 22.)

160. The earth being a globe, the sun constantly enlightens the half next to him,* while the other half is in darkness. The boundary between the enlightened and the unenlightened part is called *the circle of illumination.* When the earth is at one of the equinoxes, the sun is at the other, and the circle of illumination passes through both the poles. When the earth reaches one of the tropics, the sun being at the other, the circle of illumination cuts the earth so as to pass 23° 28′ beyond the nearer, and the same distance short of the remoter pole. These results would not be uniform, were not the earth's axis always to remain parallel to itself. The following figure will illustrate the foregoing statements.

Fig. 31.

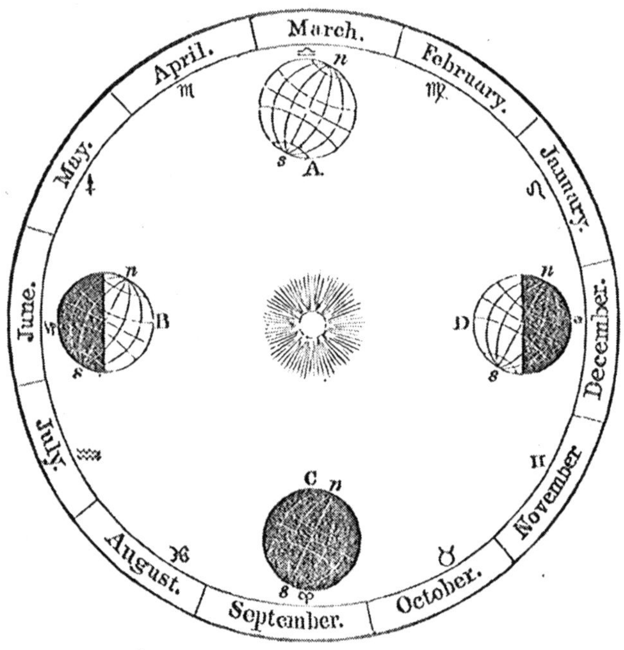

* In fact, the sun enlightens a little more than half the earth, since, on account of his vast magnitude, the tangents drawn from the sides of the sun to corresponding sides of the earth, converge to a point behind the earth, as will be seen by and by, in the representation of eclipses. The amount of illumination, also, is increased by refraction.

Let ABCD represent the earth's place in different parts of its orbit, having the sun in the center. Let A, C, be the position of the earth at the equinoxes, and B, D, its positions at the tropics, the axis *ns* being always parallel to itself.* At A and C the sun shines on both *n* and *s*; and now let the globe be turned round on its axis, and the learner will easily conceive that the sun will appear to describe the equator, which being bisected by the horizon of every place, of course the day and night will be equal in all parts of the globe.† Again, at B, when the earth is at the southern tropic, the sun shines 23½° beyond the north pole *n*, and falls the same distance short of the south pole *s*. The case is exactly reversed when the earth is at the northern tropic and the sun at the southern. While the earth is at one of the tropics, at B for example, let us conceive of it as turning on its axis, and we shall readily see that all that part of the earth which lies within the north polar circle will enjoy continual day, while that within the south polar circle will have continual night, and that all other places will have their days longer as they are nearer to the enlightened pole, and shorter as they are nearer to the unenlightened pole. This figure likewise shows the successive positions of the earth at different periods of the year, with respect to the signs, and what months correspond to particular signs. Thus the earth enters Libra and the sun Aries, on the 21st of March, and on the 21st of June the earth is just entering Capricorn and the sun Cancer.

161. Had the axis of the earth been perpendicular to the plane of the ecliptic, then the sun would always have appeared to move in the equator, the days would everywhere have been equal to the nights, and there could have been no change of seasons. On the other hand, had the inclination of the ecliptic to the equator been much greater than it is, the vicissitudes of the seasons would have been proportionally greater than at present. Suppose, for instance, the equator had been at right

* The learner will remark that the hemisphere toward *n* is above, and that toward *s* is below the plane of the paper. It is important to form a just conception of the position of the axis with respect to the plane of its orbit.

† At the pole, the solar disk, at the time of the equinox, appears bisected by the horizon.

angles to the ecliptic, in which case, the poles of the earth would have been situated in the ecliptic itself; then in different parts of the earth the appearances would have been as follows:—To a spectator on the equator, the sun, as he left the vernal equinox would every day perform his diurnal revolution in a smaller and smaller circle, until he reached the north pole, when he would halt for a moment and then wheel about and return to the equator in the reverse order. The progress of the sun through the southern signs, to the south pole, would be similar to that already described. Such would be the appearances to an inhabitant of the equatorial regions. To a spectator living in an oblique sphere, in our own latitude for example, the sun, while north of the equator, would advance continually northward, making his diurnal circuits in parallels further and further distant from the equator, until he reached the circle of perpetual apparition, after which he would climb by a spiral course to the north star, and then as rapidly return to the equator. By a similar progress southward, the sun would at length pass the circle of perpetual occultation, and for some time (which would be longer or shorter, according to the latitude of the place of observation) there would be continual night.

The great vicissitudes of heat and cold which would attend such a motion of the sun, would be wholly incompatible with the existence of either the animal or the vegetable kingdoms, and all terrestrial nature would be doomed to perpetual sterility and desolation. The happy provision which the Creator has made against such extreme vicissitudes, by confining the changes of the seasons within such narrow bounds, conspires with many other express arrangements in the economy of nature to secure the safety and comfort of the human race.

FIGURE OF THE EARTH'S ORBIT.

162. Thus far we have taken the earth's orbit as a great circle, such being the projection of it on the celestial sphere; but we now proceed to investigate its actual figure.

Were the earth's path a circle, having the sun in the center, the sun would always be at the same distance from us; that is, the *radius vector* (the name given to a line drawn from the center of the sun to the orbit of any planet) would always be of the same length. But the earth's distance from the sun is constantly varying, which shows that its orbit is not a circle, having the sun at the center. We learn the true figure of the orbit, by ascertaining the *relative distances* of the earth from the sun at various periods of the year. When these are laid down according to their relative length, and making angles with each other equal to the changes in the sun's angular motions, a curve joining the extremities of these lines gives us our first idea of the shape of the orbit, which is found to be an ellipse. Thus (considering the earth E, for the present, as fixed, and the sun as the moving body), let E*a*, E*b*, E*c*, &c. (Fig. 32), be the successive distances of the sun, laid down as just described; then will the dotted line *afmt*, which passes through their extremities, show the form of the apparent solar orbit, with the earth in one of its foci.

Fig. 32.

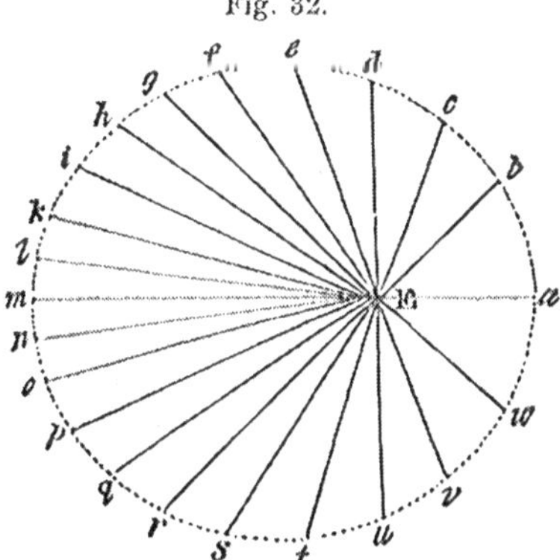

163. These relative distances may be found by observing the changes in the sun's apparent diameter. Were the variations in the sun's horizontal parallax considerable, as is the case with the moon's, this might be made the measure of the relative distances, for the parallax varies inversely as the distance (Art. 82); but the whole horizontal parallax of the sun is only 9″, and its variations are too slight and delicate, and too difficult to be found, to serve as a criterion of the changes in the sun's distance from the earth. But the changes in the *sun's apparent diameter* are much more sensible, and furnish a better method of measuring the relative distances of the earth from the sun. By a principle in optics, the apparent diameter of an object, at different distances from the spectator, is inversely as the distance.* Hence, the lines Ea, Eb, &c. (Fig. 32), are to be drawn proportional to the reciprocals of the apparent diameters of the sun.

164. The point where the earth, or any planet, in its revolution, is nearest the sun, is called its *perihelion;* the point where it is furthest from the sun, its *aphelion.* The place of the earth's perihelion is known, since there the apparent magnitude of the sun is greatest; and when the sun's magnitude is least, the earth is known to be at its aphelion. The sun's apparent diameter when greatest is 32′ 35″.6; and when least, 31′ 31″; hence the radius vector at the aphelion : rad vector at the perihelion :: 32.5933 : 31.5167 :: 1.034 : 1. Half of the difference of the two is equal to the distance of the focus of the ellipse from the center, a quantity which is always taken as the measure of the *eccentricity* of a planetary orbit. From twenty-four observations made with the greatest care by Dr. Maskelyne, at the Royal Observatory of Greenwich, the following distances of the earth from the sun are determined for each month in the year.

Time of Observation.		Distances.	Time of Observation.		Distances.
January	12–13,	0.98448	July	18–19,	1.01658
February	17–18,	0.98950	August	26–27,	1.01042
March	14–15,	0.99622	September	22–23,	1.00283
April	28–29,	1.00800	October	24–25,	0.99303
May	15–16,	1.01234	November	18–20,	0.98746
June	17–18,	1.01654	December	17–18,	0.98415

165. Having determined the form of the solar orbit, we are prepared to see what relation exists between the sun's *angular velocity* in this orbit, and the length of the radius vector. It has been already noticed (Art. 105), that the sun's progress in the ecliptic is fastest near the perihelion, and slowest near the aphelion. For instance, the sun at perihelion ad-

* More exactly, the *tangent* of the apparent diameter is inversely as the distance; but in small angles like those concerned in the present inquiry, the angle and its tangent vary alike.

vances about 61′ in 24 hours, and at aphelion only about 57′. Now $\frac{61}{57} = 1.07$, which is the square of 1.034, the ratio of apparent diameters at the same points. Indeed, a careful comparison of the sun's angular velocities, in all parts of the orbit, shows that they *vary inversely as the squares of the distances.* If changes in angular (*i. e.*, apparent) velocity were caused wholly by difference of distance, then it would vary inversely only as the first power of the distance, just as the apparent diameter does. But since the angular velocity varies inversely as the *square*, instead of the first power of the distance, the *absolute* velocity must also be greater as the distance is less, and *vice versa.* Thus we perceive, that when the sun is nearest to us, he appears to move fastest for two reasons,—first, because the same rate of motion would appear greater, if nearer to us, and secondly, because the actual motion is then greater; and each of these is in the inverse ratio of the distance.

166. It must be remembered, that this reasoning proceeds on the ground that the line of motion is everywhere at right angles to the radius vector. That this is true without sensible deviation, appears from the fact that the solar orbit is very nearly circular, with the earth at its center. If truly represented on paper, it could not be distinguished by the eye from a circle.

This relation between distances and apparent velocities having been once established, advantage may be taken of the rapid rate of change in the latter, to determine the variations in the radius vector more accurately than can be done by the apparent diameter.

Fig. 33.

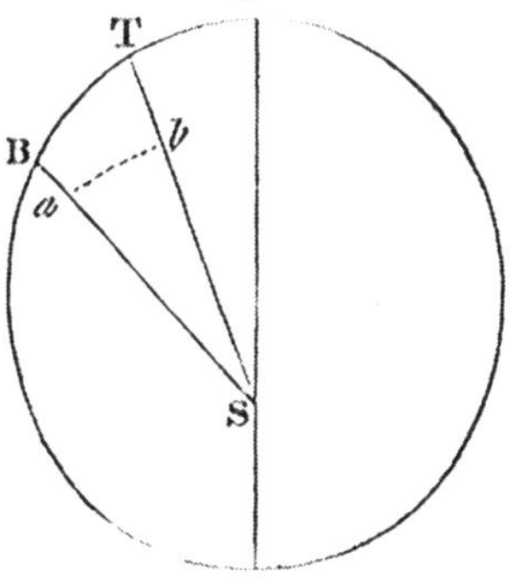

167. The angular velocity being inversely as the square of the distance in all parts of the solar orbit, it follows that the *product of the angle described in any given time, by the square of the distance, is always the same constant quantity.* For if of two factors, A×B, A is increased as B is diminished, the product of A and B is always the same. If, therefore, from the sun S (Fig. 33), two radii be drawn to T, B, the extremities of any small arc, as that described in one day, and the angle between them be called S, then $SB^2 \times S$ gives the same constant product in all parts of the orbit.

168. *The radius vector of the solar orbit describes equal areas in equal times; and in unequal times, areas proportional to the times.*

The solar orbit is so nearly a circle, that TB may everywhere be regarded as perpendicular to the radius SB or ST. Hence, the sector described in a given time, as one day, $TSB \propto SB \times TB$. But a circular arc varies both as the angle which it subtends, and also as the radius by which it is described; therefore $TB \propto SB \times S$. Hence, $TSB \propto SB^2 \times S$. But (Art. 167) this is a constant product; therefore, the area TSB is also constant, and the radius vector describes equal areas in equal times.

The sun's orbit may be accurately represented by taking some point, as the perihelion, drawing the radius vector to that point, and, considering this line as unity, drawing other radii making angles with each other such that the included areas shall be proportional to the times, and of a length required by the distance of each point as given in the table (Art. 164). On connecting these radii, we shall thus see at once how little the earth's orbit departs from a perfect circle. Small as the difference appears between the greatest and least distances, yet it amounts to nearly $\frac{1}{29}$ of the perihelion distance, a quantity no less than 3,000,000 of miles.

169. The foregoing method of determining the figure of the earth's orbit is founded on *observation;* but this figure is subject to numerous irregularities, the nature of which cannot be clearly understood without a knowledge of the leading principles of Universal Gravitation. An acquaintance with these will also be indispensable to our understanding the causes of the numerous irregularities, which (as will hereafter appear) attend the motions of the moon and planets. To the laws of universal gravitation, therefore, let us next apply our attention.

CHAPTER III.

OF CENTRAL FORCES—GRAVITATION.

170. When a body moves in a curve of any kind, we recognize the effect of *two forces:* one, an *impulse*, which acting alone would have caused a uniform motion in a straight line, and whose influence is always retained in the curve-motion; the other, an accelerating force, which continually urges the body toward some point out of the original line of motion. The first is called the *projectile* force, the other the *centripetal* force. If the action of the latter were to cease at any moment, the body by its inertia would from that moment continue uniformly in the direction in which it was then moving. Such motion in the tangent may be regarded as the effect of an impulse first given in the direction of that tangent. This supposed impulse is the projectile force for the moment in question; but it is in truth the resultant of the original impulse, and the infinite series of actions already produced by the centripetal force.

The centripetal force is of necessity infinitely small compared with the projectile force. For, if not, the curve would depart by a finite angle from the tangent; whereas, by the very nature of the relation of a curve to its tangent, the angle is infinitely small; therefore, the deflecting force is infinitely small. But it produces finite deflection after a time, because its action is incessantly repeated.

171. From a long and laborious examination of the recorded observations of Tycho Brahe, Kepler deduced three laws relating to the movements of the planets, which are therefore called Kepler's laws.

1. *The orbit of every planet is an ellipse, having the sun in one focus.*

2. *The radius vector of each orbit describes equal areas in equal times.*

3. *The squares of the periodic times of the several planets vary as the cubes of their mean distances.*

These were thus ascertained as *facts*, many years before Newton demonstrated, by mathematical reasoning, that they are necessarily involved in the laws of inertia and gravitation.

The fundamental principles of all mechanical action, pertaining alike to terrestrial bodies and to the worlds scattered throughout space, are the following:

1. Matter, until acted on by extraneous force, will remain perpetually in its present condition, whether of rest or straight uniform motion.

2. All motions communicated to a body coexist in the motion of the body.

3. To every action there is an equal and opposite reaction.

4. All masses tend toward each other, with a force varying directly as the quantity of matter, and inversely as the square of the distance.

The three first, the laws of *inertia*, of *coexistent motions*, and of *equal action and reaction*, were seen to be the true first principles in the Mechanics of terrestrial bodies. But they are equally essential in Astronomy—the *celestial* Mechanics; and not only does no fact in this science militate against them, but, on the contrary, they form the basis of all correct reasoning on the motions of the heavenly bodies. The fourth, usually called the *law of gravity*, is far more prominent in Astronomy than in Mechanics, but harmonizes with all the facts of both. We proceed to show that Kepler's laws and other laws of central forces, are the necessary consequences of the above-named mechanical principles.

172. Whatever path a body describes under the influence of a projectile and a centripetal force, *the radius vector of that path passes over equal spaces in equal times.*

Let S (Fig. 34) be the center of attraction, and suppose the projectile force in the line YR to be such as to cause the body to pass over the equal spaces PQ, QR, &c., each in a certain unit of time. When the body reaches Q, let the action toward S be sufficient to move it over QV in the same time in which by the original impulse it would describe QR. Then it will in the same time describe the diagonal QC of the parallelogram. Join RS and CS. The triangles QSC and QSR are equal; but QSR=QSP; ∴ QSC=QSP. That is, the areas de-

Fig. 34.

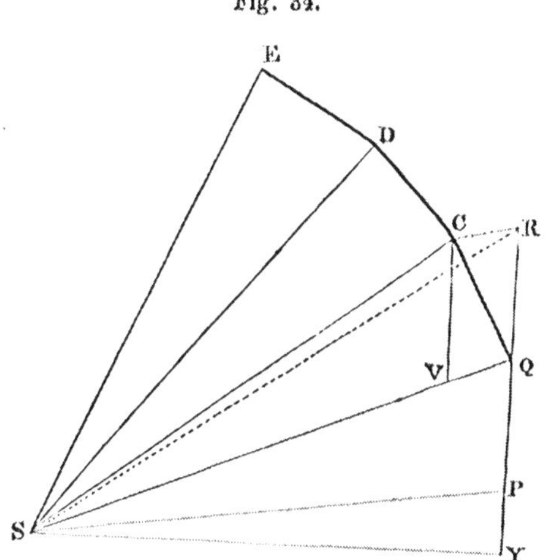

scribed in the first and second units of time are equal. In like manner, by supposing a second action toward S to occur at C, a third at D, &c., it is proved that QCS, CDS, DES, &c., which are described in equal times, are equal. This is true, however small the unit of time between the successive actions toward S, and is therefore true, when the central force acts *incessantly*, and causes curvilinear motion. As the diagonal of each parallelogram is in the same plane with its two sides, it is obvious that the whole orbit lies in one and the same plane.

173. Conversely, *if equal areas be described about a point in equal times, by the radius vector, the deflecting force acts toward that point.* For PSQ=QSR, as before (Fig. 34); and by supposition, PSQ=QSC; ∴ QSC=QSR; hence CR is parallel to QS, and QC is the diagonal of a parallelogram, whose side QV, in which the deflecting force acts, is directed toward S.

It has been shown (Art. 168) that, from observations on the angular velocity of the earth about the sun, equal areas *are* in fact described by the radius vector in equal times. It is therefore inferred that there is an accelerating force urging the earth toward the sun.

174. *The velocity at any point of an orbit, varies inversely* *as the perpendicular from the center of force to the tangent at that point.*

Let SY (Fig. 34) be perpendicular to PQ; then the area SPQ$=\frac{1}{2}$PQ×SY, which varies as PQ×SY; ∴ PQ $\propto \frac{\text{SPQ}}{\text{SY}}$. But PQ $\propto$ V, the velocity at P; and the area SPQ is constant; ∴ V$\propto \frac{1}{\text{SY}}$, or the velocity varies inversely as the perpendicular from S upon the line in which the body is moving; in other words, upon the tangent of its path, if it describes a curve.

In the orbits of the planets, since they are very nearly circular, SY meets the path almost at the point where the body is moving, and therefore is about equal to the radius vector; so that in the planetary orbits, the absolute velocity varies inversely as the radius vector very nearly. We have already noticed this to be sensibly true in the case of the earth's orbit. (Arts. 165, 166.) It follows from the above reasoning, that in a circular orbit, where the radius vector is constant, the velocity of the body is uniform.

175. When a body moves in a curve, since by its inertia it tends at each point to proceed in the tangent at that point, there is a continual outward pressure directed *from* the center of force; this is called the *centrifugal* force. It may be regarded as that infinitesimal component of the projectile force, which opposes the action of the centripetal, the motion along the curve being the other component. If the body is maintained at the same distance from the center (that is, in the circumference of a circle), the centrifugal force equals the centripetal; but in orbits of other forms, it is sometimes greater and sometimes less than the centripetal.

176. In a circular orbit, the *central force* (either centripetal or centrifugal) *varies as the square of the velocity divided by the radius.*

If v=the uniform velocity in the orbit, and t=the infinitely small portion of time of describing the minute arc Ab, and r=the radius of the circle, then Ab=vt. But Ab, or its

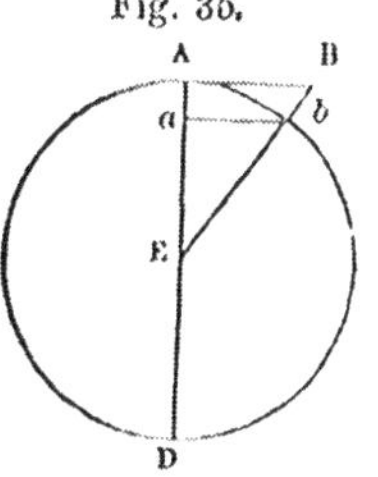

Fig. 35.

chord, is a mean proportional between its versed sine Aa, and the diameter $2r$; or $Aa = \frac{Ab^2}{2r} = \frac{v^2t^2}{2r}$, which, in a *given* time, varies as $\frac{v^2}{r}$, that is (since the central force f is measured by Aa), $f \propto \frac{v^2}{r}$.

Hence, in a given circle, where r is constant, the central force varies as the square of the velocity. In whirling a ball, for instance, with a string of given length, if the velocity is doubled, the *strain* upon the string (the centrifugal force) is four times as great, and the *strength* of the string (the centripetal force) needs also to be four times as great. If a train of cars goes round a curve with a velocity $1\frac{1}{2}$ times that which is intended, its tendency to be thrown from the track is increased $2\frac{1}{4}$ times.

177. In a circular orbit, *the central force* (centripetal or centrifugal) *varies as the radius of the circle divided by the square of the time of revolution.*

Let t=the time of describing the whole circumference $2\pi r$; $\therefore\ 2\pi r = vt$, and $v = \frac{2\pi r}{t}$, which varies as $\frac{r}{t}$; $\therefore\ v^2 \propto \frac{r^2}{t^2}$. But (Art. 176) $f \propto \frac{v^2}{r} \propto \frac{r^2}{t^2} \div r \propto \frac{r}{t^2}$; $\therefore f \propto \frac{r}{t^2}$. Hence, if the periodic time is the same, the attraction to the center must be increased in the same ratio as the radius of the orbit, for then $f \propto r$. If a string is twice as *long*, it must have twice the *strength*, in order to whirl a ball at the same rate of revolution.

178. *If a body describes an elliptical orbit by a centripetal force which acts toward the focus, that force varies inversely as the square of the distance.*

Let the body be at M (Fig. 36), and MF the radius vector at that point. Let MO be the radius of curvature at M, and therefore perpendicular to the tangent; and suppose MN to be an infinitely small arc described in a given small portion of time. Draw FP perpendicular to the tangent MP, NK to

FM, and NH to MO; then PFM, MHI, KNI, are similar triangles. MN, considered as a straight line, is described by the joint action of the centripetal force MI, and the projectile force, which is equal and parallel to IN. The motion in MI

Fig. 36.

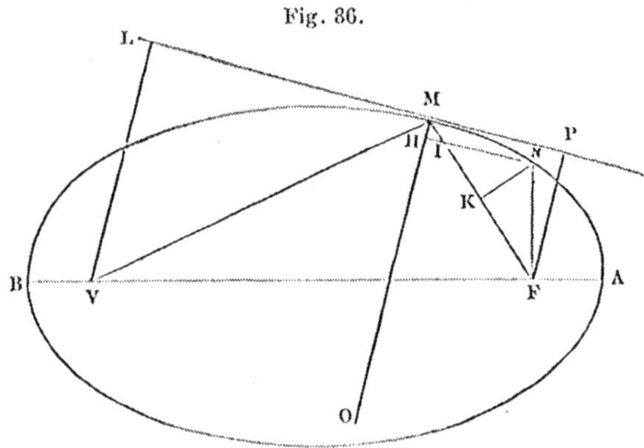

may be regarded as uniformly accelerated, because in the infinitely small time of describing it, the centripetal force may be considered constant. Hence, 2MI may be taken as the measure of the centripetal force f.* Therefore $f \propto$ MI. It is to be proved that $\text{MI} \propto \frac{1}{\text{FM}^2}$.

179. By similar triangles, MI : MH : : NI : NK;

$$\therefore \text{MI} = \text{MH}\frac{\text{NI}}{\text{NK}}.$$

Now, the chord MN is a mean proportional between the versed sine MH, and the diameter 2MO; or $\text{MH} = \frac{\text{MN}^2}{2\text{MO}}$; but, as the arc is infinitely small, NH=MN; $\therefore \text{MH} = \frac{\text{NH}^2}{2\text{MO}}$. Again, the versed sine MH, and therefore HI, is infinitely small compared with NH, and NI may be substituted for NH;

$$\therefore \text{MH} = \frac{\text{NI}^2}{2\text{MO}}.$$

* Nat. Phil., Art. 28.

Now, it is shown in conic sections, that $MO=\frac{p}{2}\left(\frac{FM}{FP}\right)^3$;* $\therefore$ by similar triangles, $MO=\frac{p}{2}\left(\frac{NI}{NK}\right)^3$. Substituting this for MO in the equation for MH above, we have $MH=\frac{NK^3}{p \cdot NI}$. Hence, in the equation for MI, we have

$$MI=\frac{NK^3}{p \cdot NI}\times\frac{NI}{NK}=\frac{1}{p}NK^2.$$

Now the sector FMN is measured by $\frac{1}{2}FM \cdot NK$; $\therefore$ $NK=\frac{2FMN}{FM}$; and $NK^2=\frac{4FMN^2}{FM^2}$; $\therefore$ $MI=\frac{4FMN^2}{p \cdot FM^2}$. But, as equal areas are described by the radius vector in equal times, FMN is constant. Therefore

$$MI\,(=f) \propto \frac{1}{FM^2};$$

that is the centripetal force varies inversely as the square of the distance.

180. It is thus proved, that in any elliptical orbit described about the focus as the center of attraction, the intensity of that attraction varies inversely as the square of the radius vector. As there is nothing in the foregoing demonstration to limit the conclusion to the orbits which are nearly circular, like those of the planets, we are at liberty to apply it to orbits of extreme eccentricity, as those of the comets. And it is proved by Newton in his Principia, that the same law of force is necessary, in order that a body may describe any one of the conic sections about its focus as the center of attraction.

181. And not only does this law prevail in all parts of any

---

* Jackson's Conic Sections. The same may be derived from Coffin's Con. Sec., Pr. V., Curvature. R^2 or $MO^2=\frac{(FM \cdot MV)^3}{a^2b^4}$ a and b being the semi-axes; $\therefore$ $MO=\frac{1}{ab}\times(FM \cdot MV)^{\frac{3}{2}}$. Multiply by $(b^2)^{\frac{3}{2}}$, and divide by its equal $(FP \cdot VL)^{\frac{3}{2}}$; then $MO=\frac{b^2}{a}\left(\frac{FM \cdot MV}{FP \cdot VL}\right)^{\frac{3}{2}}=\frac{b^2}{a}\left(\frac{FM^2}{FP^2}\right)^{\frac{3}{2}}$, since FMP and VML, are similar. But $\frac{a^2}{b}=\frac{p}{2}$, p being the parameter; $\therefore$ $MO=\frac{p}{2}\left(\frac{FM^2}{FP^2}\right)^{\frac{3}{2}}=\frac{p}{2}\left(\frac{FM}{FP}\right)^3$.

one orbit, but it is true also, that all the different bodies of a system, describing orbits about the same center of force, are urged toward that center by attractions which vary, from one orbit to another, inversely as the square of the distance.

Let a be the semi-major, and b the semi-minor axis of any elliptic orbit. Then a is the mean distance of all points of the orbit from the focus. By a rule of mensuration, the area of the ellipse$=\pi ab$. If $s=$the area described by the radius vector in a unit of time, as *one second*, and $t=$the number of seconds in the whole period of revolution, then the ellipse also$=ts$. Therefore $\pi ab=ts$; and $t=\frac{\pi ab}{s}$; and $t^2=\frac{\pi^2a^2b^2}{s^2}$. By Kepler's third law (Art. 171), $t^2 \propto a^3$; $\therefore \frac{a^2b^2}{s^2} \propto a^3$; $\therefore \frac{b^2}{a} \propto s^2$. But, because the semi-parameter $\frac{p}{2}$, is a third proportional to the semi-axes a and b, $\frac{b^2}{a}=\frac{p}{2}$; $\therefore \frac{p}{2} \propto s^2$. Hence, substituting $\frac{p}{2}$ for s^2, that is, FMN^2, in the equation for MI (Art. 179), we find $\text{MI}=\frac{4\text{FMN}^2}{p\,.\,\text{FM}^2}=\frac{2p}{p\,.\,\text{FM}^2}=\frac{2}{\text{FM}^2}$; $\therefore f \propto \frac{1}{\text{FM}^2}$. Or, the force varies inversely as the square of the distance, in different orbits, as well as in different parts of the same orbit.

The satellites which revolve about the planets, are found to conform to Kepler's laws, and therefore the force which urges them toward their respective primaries, varies in each case inversely as the square of the distance.

182. But the inquiry still remains, does the law of gravity, as demonstrated in the foregoing articles, hold good at the *smallest* distances also? For example, do the tendencies of bodies resting on the earth, and of those elevated in the air, and of the moon, toward the earth's center, come under the same general law? This is the very question which presented itself to the mind of Newton, after he had discovered that the force which deflects the planets from their lines of motion toward the sun, varies inversely as the square of their distance from it. As he noticed the fall of an apple, the inquiry arose, may not this *fall* be of the same nature as the *bending* of the moon's path toward the earth, and may not the force in the two cases be as the squares of the distances inversely?

Let us then find what is the distance through which the moon actually *descends* in a second of time. Let the earth be at E (Fig. 37), and Ab the arc described by the moon in one second. As she was going toward B at the point A, she would have gone over the line AB in one second, if some influence had not turned her aside. This influence must be directed *toward* the earth E, because it is around E that the radius vector describes equal areas in equal times (Art. 173). Therefore Bb, or the versed sine Aa (which may be considered equal to it), is the distance fallen through in one second. Now the distance of the moon from the earth's center is 238,545 miles (Art. 201). Hence, the circumference is known. The time of revolution is 27.32 days. (Art. 213.) Therefore Ab, the distance traveled in one second is obtained. This is so small, that its versed sine Aa can not be calculated by ordinary trigonometrical tables; but is easily and accurately determined by geometry; thus, 2AE : Ab : : Ab : Aa; since the arc Ab and the chord Ab may be considered identical. Aa, thus calculated, is found to be 0.0535 of an inch. At the surface of the earth, a body falls about $16\frac{1}{12}$ feet in the first second. In order to diminish this for the moon's distance, we make the proportion, $(238.545)^2 : (3956)^2 :: 16\frac{1}{12}$ ft. : 0.0536, agreeing very accurately with the distance which the moon actually falls from a tangent in a second of time. When Newton, however, first made the comparison just described, the result was quite unsatisfactory. The tendency of the moon was about $\frac{1}{6}$ too great. Near 20 years afterward, when a new measurement of a degree of the meridian had been made, and thus a corrected magnitude of the earth, he repeated the process, and found the law of attraction in this case to be the same as elsewhere in the solar system.

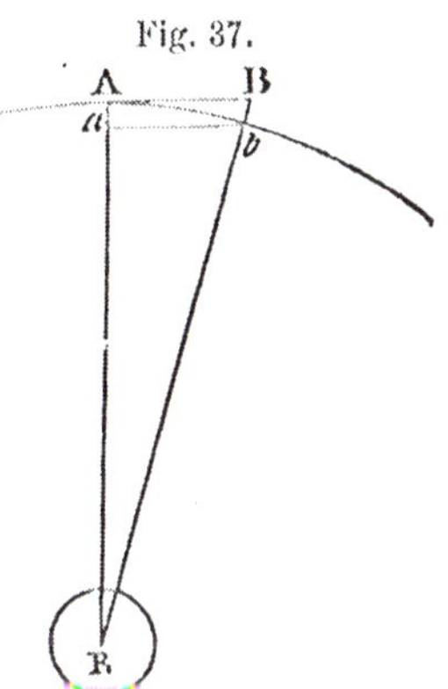

Fig. 37.

Again, the numerous disturbances which the bodies of the

solar system produce in each other's motions, are all accounted for by applying the same law. If a planet moves for a time toward another, it is accelerated; and its acceleration is greater, as the square of the distance is less, and it is retarded according to the same law, when departing from it.

Since all bodies manifest a tendency toward each other, it is natural to suppose that the tendency should vary as the quantity of matter, other things being equal. Both observation and calculation confirm, without any exceptions, such a supposition; and therefore a full statement of the law of gravity includes the fact, that it varies directly as the quantity of matter.

183. Since, as we have seen, gravity just at the earth's surface is governed by the same law of distance as it is further off, therefore the curved paths of projectiles are of the same nature as the orbits of planets and satellites; that is, they are ellipses, one of whose foci is at the earth's center. And there is no real discrepancy between this statement and that proved in Mechanics,* that the path of a projectile is a *parabola*. In that demonstration, it was assumed that the lines in which the gravitating force acts at each point of the path, are *parallel* to each other, and that the force is *constant*, neither of which is strictly true; since the verticals all meet at the center of attraction, and the intensity of gravity slightly increases as the body descends. Knowing the distance and period of the moon, it is easy to find by Kepler's third law the period of revolution in case of a given projectile, if its orbit could be completed in accordance with the law. Any force, which man could apply, would carry the perigee so little beyond the center of the earth, that the mean distance might be called one-half the radius of the earth. Therefore, calling the moon's distance 60 radii, and her period $27\frac{1}{3}$ days, we should have $(60)^3 : (\frac{1}{2})^3 :: (27\frac{1}{3})^2 : x^2$, the square root of which would be about 30 minutes. Every projectile, then, if it were free to complete its orbit, unobstructed, and according to the law of gravity which prevails *outside* of the earth, would make an entire revolution, and return to its place in half an hour. Ac-

* Nat. Phil., Art. 49.

cording to Art. 174, the velocity at the perihelion would be as much greater than that of projection as its distance from the center is less.

Fig. 38.

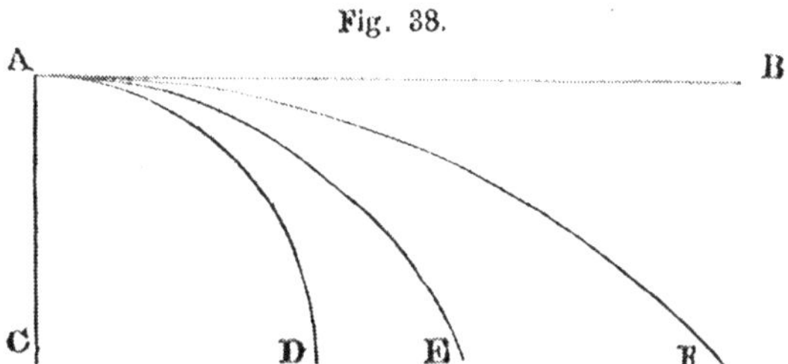

184. In Fig. 38, let AC be the vertical, and AB, at right angles to it, the line of projection; then, according to the velocity given, AD, AE, or AF might be the commencement of the orbit; from the point A the velocity would increase continually till the moment of nearest approach to the center, from which point it would decrease all the way back to A. In all these orbits, the center of the earth will occupy the focus most remote from the point of projection; in other words, the place of projection is the apogee of the orbit. Now suppose

Fig. 39.

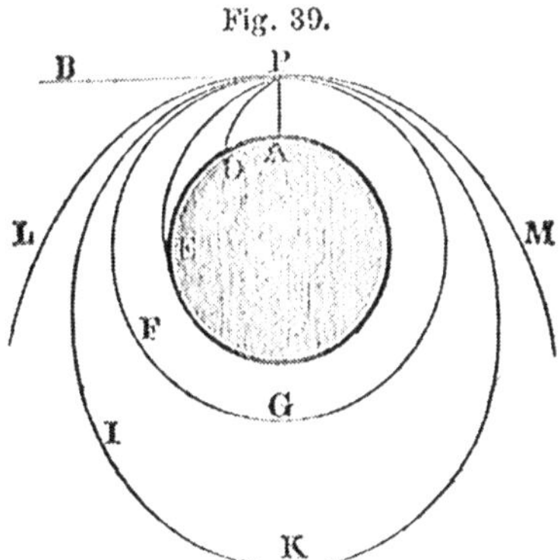

that the velocity should be greatly augmented, as in Fig. 39, so that, in one case, the curve should strike the earth at D, in another at E, and so on, until finally a force should be given sufficient to carry the body round in a circle. This last case will happen when the centrifugal force has been increased so as to just equal gravity (Art. 175). As the mean distance now equals the radius of the earth, the time of revolution is easily found to be 1h. 24m. 39s. Any increase of projectile force beyond this will again produce an ellipse as PK, whose *perigee* is at P; and we can imagine the velocity of projection increased until the ellipse becomes one of extreme eccentricity, and then changes into a parabola, and then into an hyperbola, in which last cases the body will never return, or even reach the point of apogee.

185. If we suppose the projectile motion of the earth or any other planet, to have been produced by a single impulse, that impulse may also have caused the diurnal rotation of the body. If it had been directed in a line passing through the center of gravity of the planet, then it would have caused a progressive motion, without rotation on the axis. But if the line of the impulse did not pass through the center of gravity, then besides the motion forward, there would also be a rotation whose velocity would depend on the distance of the line from the center. According to the calculation made by Bernouilli, the earth's progressive and rotary motions might have both been produced by the application of a force in a line passing twenty-four miles from the center of the earth, on the side most remote from the sun.*

Had it been directed through a point lying on the side nearest the sun, the diurnal rotation would obviously have been retrograde.

186. But if a force were applied to a planet as we have been supposing, what effect would be produced on the system as a whole? To simplify the case, suppose the sun and one planet at rest, and at a given distance apart. Let them begin to attract each other, and at the same instant let an impulse be applied to the planet at right angles to the line joining the two, and of sufficient intensity to cause revolution in a circle; will that circle be described about the sun as a stationary body? It will not, but around the common center of gravity as a mov-

* Francœur, Uran., p. 49.

ing point, while the sun will do the same in consequence of the mutual attraction between them. Figure 40 will illustrate the case. It is proved (Nat. Phil., Art. 89) that a force applied to one body of a system will produce the same effect on the center of gravity of that system, as if the force were applied to

Fig. 40

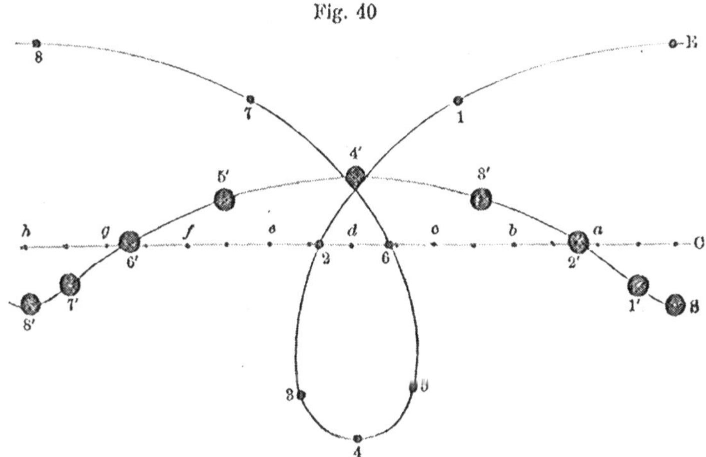

the entire system collected at that center. Therefore, let S be the sun, E the earth, and C their center of gravity; and let the impulse be such as to move the sum of both bodies over *Ca*, *ab*, *bc*, &c., in any equal times, while in each unit of time it carries E over a given arc, say 45° of its circle, then, when the center is at *a*, E is at 1, 45° from a perpendicular at *a*. But S must be on the opposite side of *a*, and at the same distance from it as from C before, because E's distance from the center remains unaltered. Therefore, by the impulse given to E, and the mutual attraction between E and S, the latter has been drawn along from S to 1′. Drawing again two circles from the center *b*, one for E, the other for S, the next positions for the bodies are 2 and 2′. While E was on the upper side of *Cd*, S was drawn toward that line, and now crosses it, and by its own inertia continues upward, although E is now below the line. In this manner the two bodies revolve about a moving center, describing *circles* relatively to that, but curves of a totally different character in space. These curves are always some variety or other of the curve called an *epicycloid*. In the case represented in the figure, the smaller body describes an epicycloid which forms a series of loops, intersecting its own path at every revolution, while the path of the heavier body is of a waving form. The body E *retrogrades* on the lower part of the loop, from 3 to 5, while S advances continually, but with unequal velocities, each body being alternately drawn forward and held back by the other.

The only way in which two separate bodies could be made to rotate about a fixed center of gravity, would be to give an equal impulse to each body, and in opposite directions. Two such forces would constitute a *couple* (Nat. Phil., Art. 57), whose effect is to produce rotation merely.

187. The learner may find some difficulty in understanding why a planet, when it reaches its aphelion C (Fig. 41), where gravity is more feeble than it had previously been, should begin from that point to *approach* the sun; and again, why, at the perihelion G, where gravity is greatest, it should begin to *depart*, instead of approaching nearer and nearer in a spiral, till it falls upon the sun. This is owing to the change in velocity, and consequently in centrifugal force. As the planet ascends through H, K, and A, the action is partly *against* its motion, and consequently retards it, till the projectile force is too small to maintain it at that distance, and it begins from C to approach. In approaching, the attraction of the sun partly conspires with its own inertia, to accelerate it, through D, E, and F; its velocity is thus increased, till the centrifugal force becomes so great at G, precisely opposite to C, that it commences once more to depart in its former path. Of course these effects could not take place unless the centrifugal force varied more rapidly than the centripetal, which is true; for it is proved that the centrifugal force in an orbit varies inversely as the *cube* of the distance, while the centripetal varies only as the *square* inversely.*

Fig. 41.

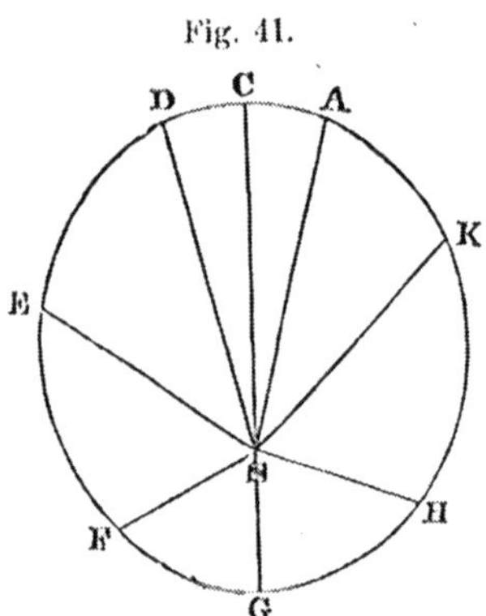

* M. Stewart's "Phys. and Math. Tracts," Prop. 8.

CHAPTER IV.

PRECESSION OF THE EQUINOXES—NUTATION—ABERRATION—MOTION OF THE APSIDES—MEAN AND TRUE PLACES OF THE SUN.

188. THE PRECESSION OF THE EQUINOXES *is a slow but continual shifting of the equinoctial points from east to west.*

Suppose that we mark the exact place in the heavens where the sun crosses the equator, the present year in March, and that this point is close to a certain star; next year the sun will cross the equator a little way westward of that star, and so every year a little further westward, until, in a long course of ages, the place of the equinox will occupy successively every part of the ecliptic, until we come round to the same star again. As, therefore, the sun, revolving from west to east in his apparent orbit, comes round toward the point where it left the equinox, it meets the equinox before it reaches that point. As the time of crossing, in every instance, *precedes* the time on the previous year, the phenomenon is called the *Precession of the Equinoxes*, and the fact is expressed by saying that the equinoxes retrograde on the ecliptic, until the line of the equinoxes makes a complete revolution from east to west. The equator is conceived as *sliding* westward on the ecliptic, always preserving the same inclination to it, as a ring placed at a small angle with another of nearly the same size, which remains fixed, may be slid quite around it, giving a corresponding motion to the two points of intersection. It must be observed, however, that this mode of conceiving of the precession of the equinoxes is purely imaginary, and is employed merely for the convenience of representation.

189. The *amount of precession* annually is 50″.1; whence, since there are 3600″ in a degree, and 360° in the whole circumference, and consequently, 1296000″, this sum divided by 50.1 gives 25868 years for the period of a complete revolution of the equinoxes.

190, Suppose now we fix to the center of each of the two rings (Art. 188) a wire representing its axis, one corresponding to the axis of the ecliptic, the other to that of the equator, the extremity of each being the pole of its circle. As the ring denoting the equator turns round on the ecliptic, which with its axis remains fixed, it is easy to conceive that the axis of the equator revolves around that of the ecliptic, and the pole of the equator around the pole of the ecliptic, and constantly at a distance equal to the inclination of the two circles. To transfer our conceptions to the celestial sphere, we may easily see that the axis of the diurnal sphere (that of the earth produced, Art. 28) would not have its pole constantly in the same place among the stars, but that this pole would perform a slow revolution around the pole of the ecliptic from east to west, completing the circuit in about 26,000 years. Hence the star which we now call the pole-star, has not always enjoyed that distinction, nor will it always enjoy it hereafter. When the earliest catalogues of the stars were made, this star was 12° from the pole. It is now 1° 24′, and will approach still nearer; or, to speak more accurately, the pole will come still nearer to this star, after which it will leave it, and successively pass by others. In about 13,000 years, the bright star Lyra, which lies on the circle of revolution opposite to the present pole-star, will be within 5° of the pole, and will constitute the Pole-star. As Lyra now passes near our zenith, the learner might suppose that the change of position of the pole among the stars, would be attended with a change of altitude of the north pole above the horizon. This mistaken idea is one of the many misapprehensions which result from the habit of considering the horizon as a fixed circle in space. However the pole might shift its position in space, we should still be at the same distance from it, and our horizon would always reach the same distance beyond it.

191. *The precession of the equinoxes is an effect of the spheroidal figure of the earth, and arises from the attraction of the sun and moon upon the excess of matter about the earth's equator.*

Were the earth a perfect sphere, the attractions of the sun and moon upon the earth would be in equilibrium among themselves. But if a globe were cut out of the earth (taking half the polar diameter for radius), it would leave a protuberant mass of matter in the equatorial regions, which may be considered as all collected into a ring resting on the earth. The sun being in the ecliptic, while the plane of this ring is inclined to the ecliptic 23° 28′, of course the action of the sun is oblique to the ring, and may be resolved into two forces, one in the plane of the equator, and the other perpendicular to it. The latter only can act as a disturbing force, and tending as it does to draw down the ring to the ecliptic, the ring would turn upon the line of the equinoxes as upon a hinge, and dragging the earth along with it, the equator would ultimately coincide with the ecliptic were it not for the revolution of the

Fig. 42.

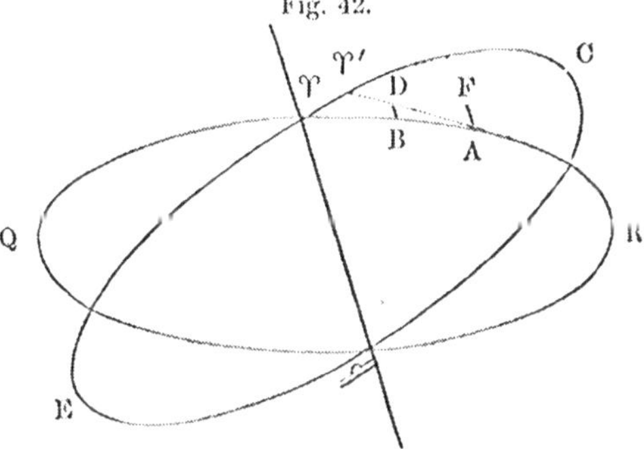

earth upon its axis. Let EC (Fig. 42) be the ecliptic, and QR the equator. Any particle A, of the ring, by its inertia of rotation, tends to move toward ♈ in the plane QR. Let AB represent this force, and AF the pressure toward EC produced by the sun; then the resultant will be AD; shifting the equinox backward from ♈ to ♈′. Each particle is subjected to this influence, except at the moment (each day) of crossing ♈ and ♎, so long as the sun is not himself in the line ♈ ♎ produced, which occurs in March and September. The effect is then interrupted for a short time.

192. The moon conspires with the sun in producing the precession of the equinoxes, its effect, on account of its nearness to the earth, being more than double that of the sun, or as 7 to 3. The planets likewise, by their attraction, produce a small effect upon the equatorial ring, but the result is slightly to diminish the amount of precession. The whole effect of the sun and moon being 50″.41, that of the planets is 0″.31, leaving the actual amount of precession 50″.1.

The law of compositions of motion in rotation is analogous to that for the composition of rectilinear motions (Nat. Phil., Art. 42), and may be stated thus: *if two forces are applied to a body, which, separately, would cause it to rotate on two different axes, their joint action will produce rotation on an axis lying between the others, at angles whose sines are inversely as the forces.* In the *precession* of *equinoxes*, the earth rotates on the diurnal axis by one force, and the sun tends to revolve it on an axis passing through the equinoxes. As the latter force is minute, compared with the other, the new axis is shifted by a very small angle each year from the axis of rotation toward the line of equinoxes. And as this line slides along by the same quantity, the two axes remain perpetually at right angles with each other.*

193. *The time occupied by the sun in passing from an equinox or solstice round to the same point again, is called the* TROPICAL YEAR. As the sun does not perform a complete revolution in this interval, but falls short of it 50″.1, the tropical year is shorter than the sidereal by 20m. 20s. in mean solar time, this being the time of describing an arc of 50″.1 in the annual revolution. The changes produced by the precession of the equinoxes in the apparent places of the circumpolar stars, have led to some interesting results in *chronology*. In consequence of the retrograde motion of the equinoctial points, the *signs* of the ecliptic (Art. 35) do not correspond at present to the *constellations* which bear the same names, but lie about one whole sign, or 30°, westward of them. Thus, that division of the ecliptic which is called the sign Taurus, lies in the constellation Aries, and the sign Gemini in the constellation Taurus. Undoubtedly, however, when the ecliptic was thus first divided, and the divisions named, the several constella-

* The precession of equinoxes, and other cases of compound rotations, are finely illustrated by the *gyroscope* of Foucault, and still better by Johnson's *rotascope*.

tions lay in the respective divisions which bear their names. How long is it, then, since our zodiac was formed?

50″.1 : 1 year : : 30° (= 108000″) : 2155.6 years.

The result indicates that the present divisions of the zodiac were made soon after the establishment of the Alexandrian school of astronomy. (Art. 2.)

NUTATION.

194. Nutation *is a vibratory motion of the earth's axis, arising from periodical fluctuations in the obliquity of the ecliptic.*

If the sun and moon moved in the plane of the equator, there would be no precession, and the effect of their action in producing it varies with their distance from that plane. Twice a year, therefore,—namely, at the equinoxes,—the effect of the sun is nothing; while at the solstices the effect of the sun is a maximum. On this account the obliquity of the ecliptic is subject to a semi-annual variation, since the sun's force, which tends to produce a change in the obliquity, is variable, while the diurnal motion of the earth, which prevents the change from taking place, is constant. Hence the plane of the equator is subject to an irregular motion, which is called the *Solar Nutation.* The name is derived from the oscillatory motion communicated by it to the earth's axis, while the pole of the equator is performing its revolution around the pole of the ecliptic (Art. 190). The effect of the sun, however, is less than that of the moon, in the ratio of 2 to 5. By the nutation alone, the pole of the earth would perform a revolution in a very small ellipse, only 18″ in diameter, the center being in the circle which the pole describes around the pole of the ecliptic; but the combined effects of precession and nutation convert the circumference of this circle into a wavy line. The motion of the equator occasioned by nutation, causes it alternately to approach to and recede from the stars, and thus to change their declinations. The solar nutation, depending on the position of the sun with respect to the equinoxes, passes through all its variations annually; but the lunar nutation, depending on the position of the moon, with respect to her nodes, varies through a period of about 18½ years.

ABERRATION.

195. Aberration *is an apparent change of place in the stars, occasioned by the joint effects of the motion of the earth in its orbit, and the progressive motion of light.*

Suppose the earth to move from C to E, while the light from S describes the line DE. If they arrive together at the point E, the impression on the eye will not be the same as if the observer had been *at rest*, but it will appear to come in the direction of S'E, the star being apparently thrown forward from S to S'. For, make EA=DE, and complete the parallelogram CA; and suppose, according to the principle of equal action and reaction, that the light has a motion EC given to it, in place of the earth's motion CE; then the two motions EA and EC will produce the resultant EB, as though the light had come from S' instead of S.

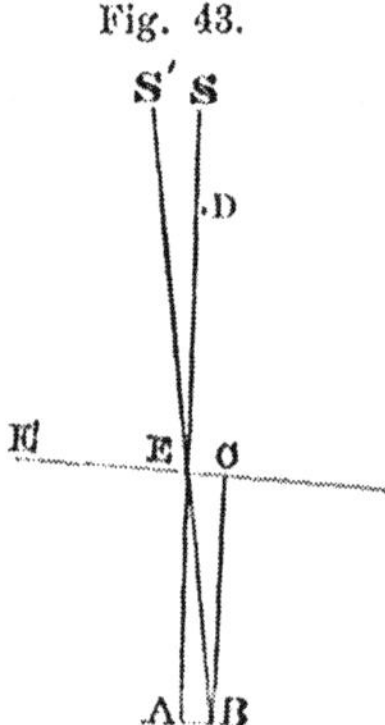

Fig. 43.

Since the earth moves 19 miles, and light 192,000 per second, if S is in a direction perpendicular to the line of the earth's motion, the right-angled triangle ECB gives about 20''.5 for the displacement of the star. In fact, however, it was the observed displacement of 20''.5 in all stars situated 90° from the direction in which the earth is moving at any time, which led to the knowledge of the velocity of light; thus,

$$\tan 20''.5 : R :: 19 : 192,000.$$

If there were no change in the aberration of a star, the fact of such aberration could never have been discovered. When we move directly *toward* or *from* a star, it plainly has no aberration; but three months before that time our motion crosses the line of the rays at right angles, and also three months after, but in a contrary direction. Hence any star in the plane of the ecliptic apparently moves back and forth over an arc of 41'' (2 × 20''.5) in a year. Those out of the ecliptic seem to describe elliptic orbits, whose major axis is 41'', and of various eccentricity, the minor axis increasing with their latitude.

MOTION OF THE APSIDES.

196. The two points of the ecliptic, where the earth is at the greatest and least distances from the sun respectively, do not always maintain the same places among the signs, but gradually shift their positions from west to east. If we accurately observe the place among the stars, where the earth is at the time of its perihelion the present year, we shall find that it will not be precisely at that point the next year when it arrives at its perihelion, but about 12″ (11″.66) to the east of it. And since the equinox itself, from which longitude is reckoned, moves in the opposite direction 50″.1 annually, the longitude of the perihelion increases every year 61″.76, or a little more than one minute. This fact is expressed by saying that the line of the apsides of the earth's orbit has a slow motion from west to east. It completes one entire revolution in its own plane in about 100,000 years (111,110).

The mean longitude of the perihelion at the commencement of the present century was 99° 30′ 5″, and of course in the ninth degree of Cancer, a little past the winter solstice. In the year 1248, the perihelion was at the place of this solstice.

The advance of apsides is caused by the attraction of the other planets. As their weight is mostly outside of the earth's orbit, their effect is to diminish the earth's tendency toward the sun. But this influence is, on the whole, greater when the earth is most distant; that is, at aphelion. Consequently the earth passes a little further onward than at the preceding revolution, before turning to approach the perihelion. Thus the aphelion advances, and the law of revolution requires that the perihelion be opposite to it; hence that advances also.

197. The angular distance of a body from its perihelion is called its *Anomaly;* and the interval between the sun's passing the point of the ecliptic corresponding to the earth's perihelion, and returning to the same point again, is called the *anomalistic year*. This period must be a little longer than the sidereal year, since, in order to complete the anomalistic revolution, the sun must traverse an arc of 11″.66 in addition to 360°. Now 360° : 365.256 : : 11″.66 : 4m. 44s.

198. Since the points of the annual orbit, where the sun is at the greatest and least distances from the earth, change their position with respect to the solstices, a slow change is occasioned in the duration of the respective seasons. For, let the perihelion correspond to the place of the winter solstice, as was the case in the year 1248; then as the sun moves more rapidly in that part of his orbit, the winter season will be shorter than the summer. But, again, let the perihelion be at the summer solstice, as it will be in the year 11740,* then the sun will move most rapidly through the summer months, and the winters will be longer than the summers. At present the perihelion is so near the winter solstice, that, the year being divided into summer and winter by the equinoxes, the six winter months are passed over between seven and eight days sooner than the summer months.

MEAN AND TRUE PLACES OF THE SUN.

199. *The Mean Motion of any body revolving in an orbit, is that which it would have if, in the same time, it revolved uniformly in a circle.*

In surveying an irregular field, it is common first to strike out some regular figure, as a square or a parallelogram, by running long lines, and disregarding many small irregularities in the boundaries of the field. By this process, we obtain an approximation to the contents of the field, although we have perhaps thrown out several small portions which belong to it, and included a number of others which do not belong to it. These being separately estimated and added to or subtracted from our first computation, we obtain the true area of the field. In a similar manner, we proceed in finding the place of a heavenly body, which moves in an orbit more or less irregular. Thus we estimate the sun's distance from the vernal equinox for every day of the year at noon, on the supposition that he moves uniformly in a circular orbit: This is the sun's *mean longitude.* We then apply to this result various corrections for the irregularity of the sun's motions, and thus obtain the *true longitude.*

200. The corrections applied to the mean motions of a heav-

* Biot.

enly body, in order to obtain its true place, are called *Equations*. Thus the elliptical form of the earth's orbit, the precession of the equinoxes, and the nutation of the earth's axis, severally affect the place of the sun in his apparent orbit, for which equations are applied. In a collection of Astronomical Tables, a large part of the whole are devoted to this object. They give us the amount of the corrections to be applied under all the circumstances and constantly varying relations in which the sun, moon, and earth are situated with respect to each other. The angular distance of the earth or any planet from its perihelion, on the supposition that it moves uniformly in a circle, is called its *Mean Anomaly:* its actual distance at the same moment in its orbit is called its *True Anomaly.*

Thus in figure 44, let AEP represent the orbit of the earth having the sun in one of the foci at S. Upon AP describe the circle AMP. Let E be the place of the earth in its orbit, and M the corresponding place in the circle; then the angle MCP is the mean, and ESP the true anomaly. The difference between the mean and true anomaly, ESP − MCP, is called *the Equation of the Center*, being that correction which depends on the elliptical form of the orbit, or on the distance of the center of attraction from the center of the figure, that is, on the eccentricity of the orbit. It is much the greatest of all the corrections used in finding the sun's true longitude, amounting, at its maximum, to nearly two degrees (1° 55′ 26″.8).

Fig. 44.

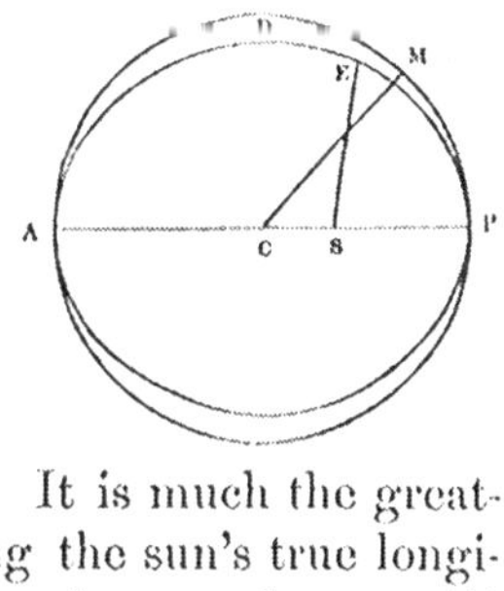

Considering the mean and true anomaly as agreeing at P, the true place of the earth E is in advance of the mean place M, because the velocity near the perihelion is greater than the mean velocity. This difference between the mean and true places (equation of the center), increases till the earth has advanced about 90° to D, when the velocity has reached its mean value; from D to A, the true place moves slower than the mean, until at A the gain and loss are balanced, and the true and mean again coincide. But, as the earth's motion is now slower than the average, the true place falls behind the mean, and the equation is negative, till the earth returns to the perihelion.

CHAPTER V.

OF THE MOON—LUNAR GEOGRAPHY*—PHASES OF THE MOON—HER REVOLUTIONS.

201. NEXT to the Sun, the Moon naturally claims our attention.

The Moon is an attendant or satellite to the earth, around which she revolves at the distance of nearly 240,000 miles. Her mean horizontal parallax being 57′ 09″,† consequently, sin 57′ 09″ : semi-diameter of the earth (3956.2) : : rad : 238,545. (Art. 87.)

The moon's *apparent diameter* is 31′ 7″, and her *real diameter* 2160 miles. For,

Rad : 238,545 : : sin 15′ 33½″ : 1079.6.=moon's semi-diameter. (See Fig. 26, p. 71.)

And, since spheres are as the cubes of the diameters, the *volume* of the moon is $\frac{1}{49}$ that of the earth. Her *density* is nearly $\frac{3}{5}$ (.615) the density of the earth, and her *mass* (=$\frac{1}{49}\times$ .615) is about $\frac{1}{80}$.

202. The moon shines by *reflected light* borrowed from the sun, and, when full, exhibits a disk of silvery brightness, diversified by extensive portions partially shaded. By the aid of the telescope, we see undoubted signs of a varied surface, composed of extensive tracts of level country, and numerous mountains and valleys.

203. The line which separates the enlightened from the dark portions of the moon's disk, is called the *Terminator*. (See Fig. 2, *Frontispiece*.) As the terminator traverses the disk from new to full moon, it appears through the telescope exceedingly broken in some parts, but smooth in others, indicating that some portions of the lunar surface are uneven while

* *Selenography* is a word more appropriate to a description of the moon, but is not perhaps sufficiently familiarized by use.

† Baily's Astronomical Tables.

others are level. The broken regions appear brighter than the smooth tracts. The latter were formerly taken for seas, and received names accordingly; as A (Fig. 2, Fr.), *mare humorum;* B, *mare nubium*, &c. But improved telescopes have shown that they are extensive plains, having inequalities which, though comparatively low, are permanent. That there are *mountains*, is known by the following indications. As the terminator advances over the disk, the light strikes the highest peaks, which appear as bright points a little way upon the dark part of the moon. After the terminator has passed over them, they project shadows away from the sun, corresponding to the apparent shape of the mountain, and growing shorter, as the rays fall more nearly vertical. And again, in the waning of the moon, the shadows are cast in the opposite direction, lengthening until the dark part of the disk reaches them, and the summits once more become isolated bright points, and then disappear. A view with a good telescope, continued for fifteen minutes, will often show perceptible changes in the position of the shadows, and the shape of illuminated peaks.

204. The *valleys* very generally have a circular form, varying in diameter from a mile or two up to sixty miles; the mountain ridge, which surrounds them, being of a ring form, generally much more precipitous on the inner side than the outer. The heights surrounding these valleys are called *ring-mountains.* One or more conical mountains frequently occupy the center of the area inclosed within the ring. Some of the principal are, No. 1. *Tycho;* 2. *Kepler;* 3. *Copernicus;* 4. *Aristarchus*, &c. (Fig. 1, Fr.) There are also larger and less regular areas, surrounded by mountains, called *bulwark plains.* The diameter of some is 130 miles; and they generally have small mountains, both of the conical and ring form, scattered over the plain. The largest are *Clavius*, *Walther*, *Regiomontanus*, *Parbuch*, *Alphonse*, *Ptolemæus.* Besides the peculiar forms now mentioned, there are chains and spurs of mountains, resembling terrestrial ranges.*

* The lunar map of Beer and Madler, 2½ feet in diameter, contains a very perfect delineation of the mountains and valleys of the moon, accompanied by their names.

205. The appearances described in the foregoing articles are obviously due to differences of elevation, since they are revealed by the sun-light falling on them at considerable obliquity. They come into view in succession, as the sun rises upon them, during the first two quarters, or sets and leaves them in shade in the last two. At the full moon, the sun's rays and our line of vision coincide in direction, and no shadows appear, and the features of mountain and valley are only imperfectly seen. But a peculiarity of another kind presents itself at the time of full moon. There are luminous stripes extending from several of the ring-mountains in straight lines to the distance of hundreds of miles. They are not ridges, since they cast no shadows as the terminator passes them; and the difference of illumination must result from difference of reflective power. But perhaps no satisfactory reason can be given for their remarkable arrangement. They are sometimes termed *lava lines*. The most extensive systems of this kind occur around Tycho, Copernicus, Kepler, and Anaxagoras.

There is no appearance of fluidity on the lunar surface, nor any such condition as we might expect to result from the flow of a liquid. All inequalities are angular and rigid, instead of being softened down by the action of water. There are no changes which might be ascribed to the growth and decay of vegetation. No spot is ever concealed by a cloud, or dimmed by an impure atmosphere. The cold is probably too intense for the existence of a liquid substance, just as the earth would be, if it were destitute of an atmosphere, and thus exposed to the low temperature of the space through which it travels.*

206. *The method of estimating the height of lunar mountains is as follows:*

Let ABO (Fig. 45) be the illuminated hemisphere of the moon, SO a solar ray touching the moon in O, a point in the circle which separates the enlightened from the dark part of the moon. All the part ODA will be in darkness; but if this part contains a mountain MF, so elevated that its summit M reaches to the solar ray SOM, the point M will be enlightened.

* For a fuller description of the moon's surface, see Lardner's "Hand-book on Meteorology and Astronomy," pp. 208–212.

Let E be the place of an observer on the earth, and suppose a line ES to be drawn to the sun; then MES is the elongation of the point M from the sun, which is determined by observation. The angle EMS can never differ so much as 9′ from the

Fig. 45.

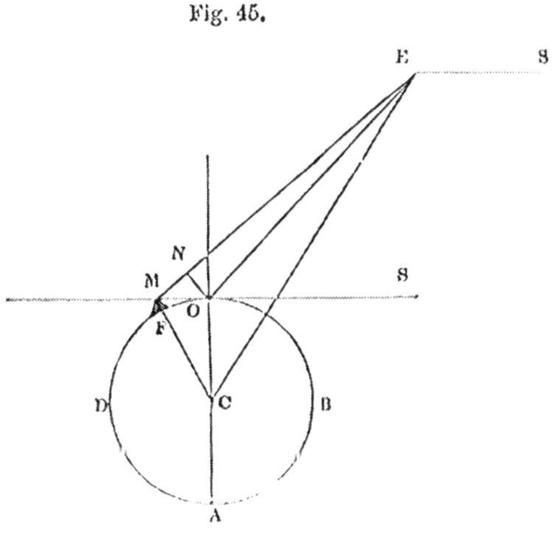

supplement of MES, on account of the great distance of the sun compared with EM, which is about 400 to 1. But to obtain EMS accurately, 400 : 1 : : sin MES : sin ESM, which angle subtracted from the supplement of MES, leaves EMS. Let MEO, the visual angle between the mountain top and the terminator, be measured by a micrometer. Then, as the distance from the earth to the moon is known, the isosceles triangle NEO gives the length of ON. As E is very small, the angles at N may be considered right angles; hence, the right-angled triangle OMN, in which ON and M are known, gives OM. Finally, from the known sides OC and OM, in the right-angled triangle OMC, CM is obtained; and OC being subtracted from it, we find MF, the height of the mountain.

207. As OM is very small compared with OC, FM is both more *easily* and more *accurately* found by taking the third proportional to 2OC and OM. For, suppose a perpendicular drawn from F to OC; then OM may be called equal to FO, and FM to the versed sine; hence 2OC : MO : : MO : FM. The whole work may be finished by logarithms, without taking from the tables any natural number, except FM the quantity sought. The heights of some mountains, determined in this way, have been found between three and four miles.

When the moon is exactly at quadrature, EMS is a right angle, and OM (being coincident with ON) is obtained directly from the micrometrical measurement of the angle E; from which FM is derived as before.

208. Schroeter, a German astronomer, estimated the heights of the lunar mountains by observations on their *shadows*. He made them in some cases as high as $\frac{1}{214}$ of the semi-diameter of the moon, that is, about 5 miles. The same astronomer also estimates the depths of some of the lunar valleys at more than four miles. Hence it is inferred that the moon's surface is more broken and irregular than that of the earth, its mountains being higher and its valleys deeper in proportion to the size of the moon than those of the earth.

209. It has sometimes been supposed that there are slight indications of an atmosphere about the moon. This is probably an error. The severest test of a perceptible atmosphere would be the effect on a star, at the beginning and end of its occultation by the moon. The star would appear to be *detained* a little in its diurnal motion, just before disappearing, and just after reappearing, in consequence of the bending of the rays which come from it, as they pass the edge of the moon's disk, and probably some *loss* of light would in that case be perceptible at the same moments. But the most careful observations have failed to show any such detention; and as to loss of light, the star, on coming up to the edge of the moon disappears all at once, with a suddenness which is startling. If there is an atmosphere, it cannot have a thousandth part of the density of the earth's.*

210. The improbability of our ever identifying *artificial structures* in the moon may be inferred from the fact that a line one mile in length in the moon subtends an angle at the eye of only about one second. If, therefore, works of art were

* Lardner.

to have a sufficient horizontal extent to become visible, they can hardly be supposed to attain the necessary elevation, when we reflect that the height of the great pyramid of Egypt is less than the sixth part of a mile.

PHASES OF THE MOON.

211. The changes of the moon, commonly called her Phases, arise from different portions of her illuminated side being turned toward the earth at different times. When the moon is first seen after the setting sun, her form is that of a bright crescent, on the side of the disk next to the sun, while the other portions of the disk shine with a feeble light, reflected to the moon from the earth. Every night we observe the moon to be further and further eastward of the sun, and at the same time the crescent enlarges, until, when it has reached an elongation from the sun of nearly 90°, half her visible disk is enlightened, and she is said to be in her *first quarter*. The terminator, or line which separates the illuminated from the dark part of the moon, is convex toward the sun from the new moon to the first quarter, and the moon is said to be *horned*. The extremities of the crescent are called *cusps*. At the first quarter, the terminator becomes a straight line, coinciding with a diameter of the disk; but after passing this point, the terminator becomes concave toward the sun, bounding that side of the moon by an elliptical curve, when the moon is said to be *gibbous*. When the moon arrives at the distance of 180° from the sun, the entire circle is illuminated, and the moon is full. She is then *in opposition* to the sun, rising about the time the sun sets. For a week after the full, the moon appears gibbous again, until, at a little less than 90° from the sun, she resumes the same form as at the first quarter, being then at her *third quarter*. From this time until new moon, she exhibits again the form of a crescent before the rising sun, until approaching her *conjunction* with the sun, her narrow thread of light is lost in the solar blaze; and finally, at the moment of passing the sun, the dark side is wholly turned toward us and for some time we lose sight of the moon.

The two points in the orbit corresponding to new and full moon respectively, are called by the common name of *syzygies;*

those which are 90° degrees from the sun are called *quadratures;* and the points half way between the syzygies and quadratures are called *octants*. The circle which divides the enlightened from the unenlightened hemisphere of the moon, is called the *circle of illumination:* that which divides the hemisphere that is turned toward us from the opposite one is called the *circle of the disk*. The degree of each phase depends on the angle between these circles.

212. As the moon is an opaque body of a spherical figure, and borrows her light from the sun, it is obvious that that half only which is toward the sun can be illuminated. More or less of this side is turned toward the earth, according as the moon is at a greater or less elongation from the sun. The reason of the different phases will be best understood from a

Fig. 46.

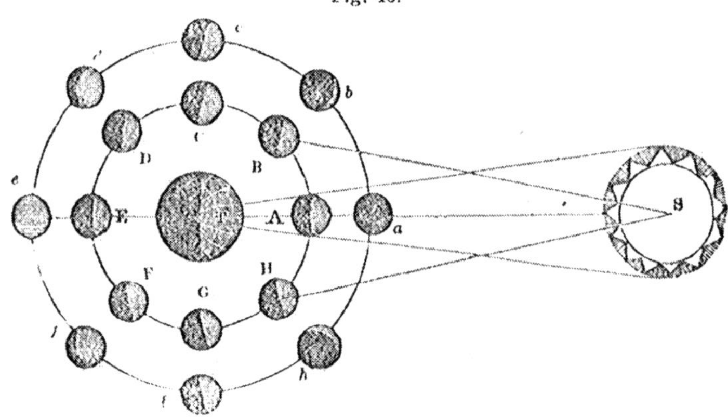

diagram. Therefore, let T (Fig. 46) represent the earth, and S the sun. Let A, B, C, &c., be successive positions of the moon. At A the entire dark side of the moon being turned toward the earth, the disk would be wholly invisible. At B, the circle of the disk cuts off a small part of the enlightened hemisphere, which appears in the heavens at *b*, under the form of a crescent. At C, the first quarter, the circle of the disk cuts off half the enlightened hemisphere, and the moon appears *dichotomized* at *c*. In like manner it will be seen that the appearances presented at D, E, F, &c., must be those represented at *d*, *e*, *f*.

REVOLUTIONS OF THE MOON.

213. *The moon revolves around the earth from west to east, making the entire circuit of the heavens in about $27\frac{1}{4}$ days.*

The precise law of the moon's motions in her revolution around the earth is ascertained, as in the case of the sun (Art. 155), by daily observations on her meridian altitude and right ascension. Thence are deduced by calculation her latitude and longitude, from which we find, that the moon describes on the celestial sphere a great circle of which the earth is the center.

The period of the moon's revolution from any point in the heavens round to the same point again, is called a *month*. A *sidereal month* is the time of the moon's passing from any star, until it returns to the same star again. A *synodical month** is the time from one conjunction or new moon to another. The synodical month is about $29\frac{1}{2}$ days, or more exactly, 29d. 12h. 44m. 2ˢ.8=29.53 days. The sidereal month is about two days shorter, being 27d. 7h. 43m. 11ˢ.5=27.32 days. As the sun and moon are both revolving in the same direction, and the sun is moving nearly a degree a day, during the 27 days of the moon's revolution, the sun must have moved 27°. Now since the moon passes over 360° in 27.32 days, her daily motion must be 13° 17′. It must therefore evidently take about two days for the moon to overtake the sun. The difference between these two periods may, however, be determined with great exactness. The middle of an eclipse of the sun marks the time of conjunction or new moon; and by dividing the interval between any two distant solar eclipses by the number of revolutions of the moon, or *lunations*, we obtain the precise period of the synodical month. Suppose, for example, two eclipses occur at an interval of 1,000 lunations; then the whole number of days and parts of a day that compose the interval divided by 1,000 will give the exact time of one lunation.† The time of the synodical month being ascertained, the exact period of the sidereal month may be derived from it. For the

---

* συν and οδος, implying that the two bodies *come together*.

† It might at first view seem necessary to know the period of one lunation before we could know the number of lunations in any given interval. This period is known very nearly from the interval between one new moon and another.

arc which the moon describes in order to come into conjunction with the sun, exceeds 360° by the space which the sun has passed over since the preceding conjunction, that is, in 29.53 days. Therefore,

365.24 : 360° :: 29.53 : 29°.1 = arc which the moon must describe more than 360° in order to overtake the sun. Hence, as the whole distance the moon must move from the sun to reach it again, is to one revolution of the moon, so is the *time* of accomplishing the former, to the *time* of a revolution; *i. e.*, 360°+29°.1 : 360° :: 29.53d. : 27.32d.

214. *The moon's orbit is inclined to the ecliptic at an angle of about* 5° (5° 8′ 48″). The intersections are called the *ascending* and *descending nodes:* through the ascending, the moon passes from south to north; through the descending, from north to south. The angle is found by measuring the moon's greatest latitude, which is, of course, equal to the inclination of the circles.

215. The moon, at the same age, crosses the meridian at different altitudes at different seasons of the year. The full moon, for example, will appear much further in the south when on the meridian at one period of the year than at another. This is owing to the fact that the moon's path is differently situated with respect to the *horizon*, at a given time of night, at different seasons of the year. By taking the ecliptic on an artificial globe to represent the moon's path (which is always near it, Art. 214), and recollecting that the new moon is seen in the same part of the heavens with the sun, and the full moon in the opposite part of the heavens from the sun, we shall readily see that in the winter the new moons must run low because the sun does, and for a similar reason the full moons must run high. It is equally apparent that, in summer, when the sun runs high, the new moons must cross the meridian at a high, and the full moons at a low altitude. This arrangement gives us a great advantage in respect to the amount of light received from the moon; since the full moon is longest above the horizon during the long nights of winter, when her presence is most needed. This circumstance is especially favorable to the inhabitants of the polar regions, the moon, when full, traversing that part of her orbit which lies north of the equator, and of course above the horizon of the north pole, and traversing the portion that lies south of the equator, and below the polar horizon, when new. During the polar winter, therefore, the moon, from the first to the last quarter, is commonly above the horizon, while the sun is absent; whereas, during summer, while the sun is present, the moon is above the horizon while describing her first and last quadrants.

216. About the time of the autumnal equinox, the moon when near the full, rises about sunset for a number of nights in succession; and as this is, in England, the period of harvest,* the phenomenon is called the *Harvest Moon*. To understand the reason of this, since the moon is never far from the ecliptic, we will suppose her progress to be in the ecliptic. If the moon moved in the equator, then, since this great circle is at right angles to the axis of the earth, all parts of it, as the earth revolves, would cut the horizon at the same constant angle. But the moon's orbit, or the ecliptic, which is here taken to represent it, being oblique to the equator, cuts the horizon at different angles in different parts, as will easily be seen by reference to an artificial globe. When the first of Aries, or vernal equinox, is in the eastern horizon, it will be seen that the ecliptic (and consequently the moon's orbit) makes its least angle with the horizon. Now at the autumnal equinox, the sun being in Libra, the moon at the full is in Aries, and rises when the sun sets. On the following evening, although she has advanced in her orbit about 13° (Art. 213), yet her progress being oblique to the horizon, and at a small angle with it, she will be found at this time but a little way below the horizon, compared with the point where she was at sunset the preceding evening. She therefore rises but a little later each evening than she did on the evening previous, but her *place* of rising moves rapidly northward. It should be observed, that in making her revolution round the earth, the moon must pass the first of Aries, and therefore make these small differences in the time of rising, every month. But as the moon is not *full* at the same time, except in autumn, the circumstance attracts no attention.

When the ascending node is at the vernal equinox, the angle of the moon's path with the horizon at the time of rising, is 10° smaller than when the descending node is there; and the intervals vary accordingly. This occurs once in about 18 years.

217. *The moon is about $\frac{1}{60}$ nearer to us when near the zenith than when in the horizon.*

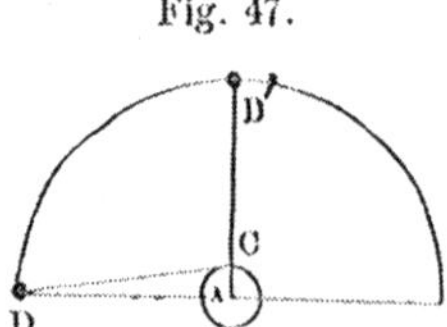

Fig. 47.

The horizontal distance CD (Fig 47) is nearly equal to AD=AD′, which is greater than CD′ by AC, the semi-diameter of the earth=$\frac{1}{60}$ the distance of the moon. Accordingly, the apparent diameter of the moon, when actually measured, is about 30″ (which equals about $\frac{1}{60}$ of the whole) greater when in the zenith than in the horizon. The apparent enlargement of the full moon when rising, is owing to the same causes as that of the sun, as explained in article 96.

218. *The moon turns on its axis in the same time in which it revolves around the earth.*

This is known by the moon's always keeping nearly the same face toward us, as is indicated by the telescope, which could not happen unless her revolution on her axis kept pace with her motion in her orbit. Thus, it will be seen by inspecting figure 31, that the earth turns different faces toward the sun at different times; and if a ball having one hemisphere white and the other black be carried around a lamp, it will easily be seen that it cannot present the same face constantly toward the lamp unless it turns once on its axis while performing its revolution. But though the same side of the moon *on the whole* is always toward the earth, yet there are small apparent oscillations, by which narrow portions of the remote side are presented alternately to view. These are called *librations*.

219. One is the *libration in longitude*, because it brings into view portions of the equator on one side and then on the other. It is owing to the fact that the moon revolves uniformly on its axis, and with unequal angular motion around the earth. Near the apogee, where she advances slowest, she describes less than 90° of her orbit while she turns just one-fourth round upon her axis; consequently showing us a little of the further side on the east limb. But in the perigeal part of her orbit, she advances faster than the mean, and therefore in one-fourth of her diurnal rotation she moves forward more than 90° in her orbit, and presents some surface beyond the west limb.

The *libration in latitude*, by which she alternately presents to our view the space about her poles, is caused by the obliquity of her equator to her orbit. Her equator is inclined about $1\frac{1}{2}°$ (1° 30′ 11″) to the *ecliptic*, and remains parallel to itself. But the angle between her equator and *orbit* varies from about $3\frac{1}{2}°$ to $6\frac{1}{2}°$, on account of the motion of the nodes. Each pole of the moon is presented toward the earth every 27 days, just as the earth's poles are turned toward the sun every year, though in a much less degree. (See Fig. 31.)

The moon exhibits another phenomenon of this kind, called her *diurnal libration*, depending on the daily rotation of the spectator. She turns the same face toward the center of the earth only, whereas we view her from the surface. When she is on the meridian, we see her disk nearly as though we viewed it from the center of the earth, and hence in this situation it is subject to little change; but when near the horizon, our circle of vision takes in more of the upper limb than would be presented to a spectator at the center of the earth. Hence, from this cause, we see a portion of one limb while the moon is rising, which is gradually lost sight of, and we see a portion of the opposite limb as the moon declines toward the west. It will be remarked that neither of the foregoing changes implies any actual motion in the moon, but that each arises from a change of position in the spectator relative to the moon.

220. An inhabitant of the moon would have but one day and one night during the whole lunar month of $29\frac{1}{2}$ days. One of its days, therefore, is equal to nearly 30 of ours. So protracted an exposure to the sun's rays, if the moon had an atmosphere like that of the earth, would occasion an excessive accumulation of heat; and so long an absence of the sun must occasion a corresponding degree of cold. Each day would be a wearisome summer; each night a severe winter.* A spectator on the side of the moon which is opposite to us would never see the earth; but one on the side next to us would see the earth presenting a gradual succession of changes during his long night of 360 hours. Soon after the earth's conjunction with the sun, he would have the light of the earth reflected to him, presenting at first a crescent, but enlarging, as the earth approaches its opposition, to a great orb, 13 times as large as the full moon appears to us, and affording nearly 13 times as much light. Our seas, plains, mountains, and clouds, would present a great diversity of appearance, as the earth performed its diurnal rotation; though the distinctness would be much impaired by the strong light reflected by our atmosphere. The earth to his view would *remain always in the same part of the sky*, having only small monthly oscillations, north and south, by means of the libration in latitude, also east and west, by the libration in longitude. For, being unconscious of his own motion around the earth, it would seem to revolve about his planet from west to east; but, meanwhile, his own diurnal rotation would give the earth an apparent motion to the west at the same mean rate, and the two would balance each other, except so far as the librations would affect them. An observer on the center of the moon's disk, would see the earth always over head; one at the edge of the disk, would see it at the horizon. The earth is full to the moon when the latter is new to us; and universally the two phases are complementary to each other.

221. If the ecliptic (the earth's path about the sun), and the moon's path about the earth, were visible lines in the sky, since we are in the plane of each, they would both appear as great circles intersecting each other in opposite points, and inclined about 5° to each other. But if we could take a view of these orbits from a distant position out of their planes, they would appear as two very unequal circles, one having a diameter 400 times greater than the other, and the small circle moving around with its center upon the circumference of the large one, once in a year.

* Francœur, Uranog., p. 91.

But again, if we confine our attention to the moon's path in its relations to the *sun*, instead of the earth, and trace it as a visible line in the solar system, we shall see that it loses entirely its character of a small circle on the circumference of a large one. On the other hand, it can hardly be distinguished from the earth's orbit itself, making 25 slight undulations alternately inside and outside of it, and never deviating from it more than one 400th part of its radius, as represented in Fig. 47′.

Fig. 47′.

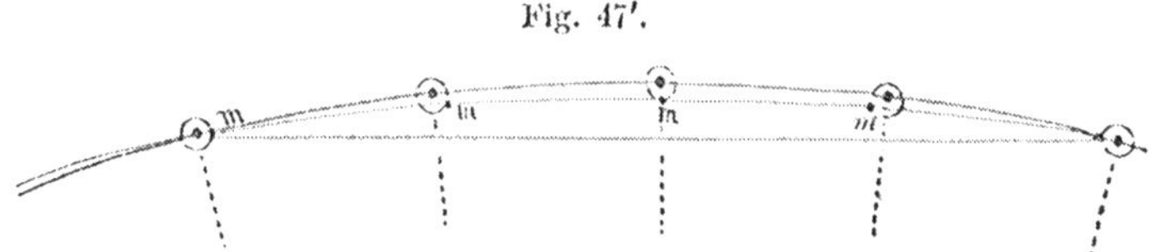

If we regard now the *forces* to which the moon is subjected in describing this path, it is clear, that since it differs so little from that of the earth, the moon must be controlled by nearly the same forces as those which keep the earth in its orbit; so that, if the earth were annihilated, the moon would still preserve its course round the sun, with so little change from its present orbit, that it would hardly be noticed by an observer who could take the whole into view at once. According to the second law of motion (Nat. Phil., Art. 12), the moon's motion round the earth simply co-exists with its motion round the sun; and the joint effect is a curve of that species called an epicycloid.

222. The relative attraction exerted by the sun and earth upon the moon, is found by applying the formula (Art. 177) $f \propto \frac{r}{t^2}$. Calling the radius of the moon's orbit=1, that of the earth's is about 400; and the times are 27.32 and 365.25. Hence, attraction to the sun : that to the earth : : $\frac{400}{(365.25)^2}$: $\frac{1}{(27.32)^2}$: : 2.2 : 1. So that the sun, though so very far from the moon, exerts upon it $2\frac{1}{5}$ times more attraction than the earth does. The moon, therefore, is much more under the influence of the sun than of the earth; the latter only causing it to oscillate on her own path as already shown. When the moon is in conjunction, the attraction of the earth diminishes its tendency to the sun, and therefore its distance increases, till it comes to the opposition; and while thus *ascending*, being in the rear of the earth, it is accelerated by it, and goes past it at the moment of opposition. But now the attraction of the earth conspires with that of the sun, so that the moon can not keep in a circular orbit at that distance, and therefore *descends* again toward the conjunction. While descending, being now in advance of the earth, it is retarded by it, and loses so much of its velocity, that on reaching the conjunction, the earth passes it again as before. Thus the moon, by the earth's influence, approaches the sun, and then recedes from it, and also gains velocity, and then loses it, suffering these changes as slight disturbances in its great annual revolution about the sun.

The earth is also affected in precisely the same manner, though in a far less degree, by the moon; it is the center of gravity of the two, which describes the annual elliptical orbit.

223. If a chord be supposed drawn in the earth's orbit (Fig. 47), between successive points of quadrature, it is found that this chord, at its middle point, falls about 600,000 miles within the orbit, while the moon is only 238,000 miles within it at the conjunction. Her path is, therefore, so near that of the earth as to be always concave toward the sun.

CHAPTER VI.

LUNAR IRREGULARITIES.

224. Considering the moon's motion simply in its relation to the earth, we have thus far spoken of its path as an ellipse, with the earth in one focus. But careful observations have proved that this elliptical revolution is subject to numerous irregularities. The law of universal gravitation has been applied with wonderful success to their investigation, and its results have conspired, with those of long-continued observation, to furnish the means of ascertaining, with great exactness, the place of the moon in the heavens at any given instant of time, past or future, and thus to enable astronomers to determine longitudes, to calculate eclipses, and to solve various other problems of the highest interest. A complete understanding of all the irregularities of the moon's motions must be sought for in more extensive treatises of astronomy than the present; but some general acquaintance with the subject, clear and intelligible, as far as it goes, may be acquired by first gaining a distinct idea of the mutual actions of the sun, the moon, and the earth.

225. *The irregularities of the moon's motions are due chiefly to the disturbing influence of the sun, which operates in two ways: first, by acting unequally on the earth and moon, and, secondly, by acting obliquely on the moon.*

If the sun acted equally on the earth and moon, and always in parallel lines, this action would serve only to restrain them in their annual motions round the sun, and would not affect their actions on each other, or their motions about their common center of gravity. In that case, if they were allowed to fall directly toward the sun, they would fall equally, and their respective situations would not be affected by their descending equally toward it. We might then conceive them as in a plane, every part of which being equally acted on by the sun, the whole plane would descend toward the sun, but the respective motions of the earth and the moon in this plane would be the same as if it were quiescent. Supposing, then, this plane and all in it to have an annual motion imprinted on it, it would move regularly round the sun, while the earth and moon would move in it, with respect to each other, as if the plane were at rest, without any irregularities. But because the moon is nearer the sun in one half of her orbit than the earth is, and in the other half of her orbit is at a greater distance than the earth from the sun, while the power of gravity is always greater at a less distance, it follows, that in one-half of her orbit the moon is more attracted than the earth toward the sun, and in the other half less attracted than the earth. The *excess* of the attraction, in the first case, and the *defect* in the second, constitutes a disturbing force; to which we may add another, namely, that arising from the *oblique action* of the solar force, since this action is not directed in parallel lines, but in lines that meet in the center of the sun; and one part of this oblique action is *in* the orbit of the moon, being now east of the earth, and then west of it; and the other part is *from* the orbit toward the plane of the ecliptic.

226. To see the effects of this process, let us suppose that the projectile motions of the earth and moon were destroyed, and that they were allowed to fall freely toward the sun. If the moon was in conjunction with the sun, or in that part of her orbit which is nearest to him, the moon would be more attracted than the earth, and fall with greater velocity toward the sun; so that the distance of the moon from the earth would be increased in the fall. If the moon was in opposition, or in the part of her orbit which is furthest from the sun, she would be less attracted than the earth by the sun, and would fall with a less velocity toward the sun, and would be left behind; so that the distance of the moon from the earth would be increased in this case also. If the moon was in one of the quarters, then the earth and moon being both attracted toward the center of the sun, they would both descend directly toward that center, and by approaching it, they would necessarily, at the same time, approach each other, and in this case their distance from each other would be diminished. Now whenever the action of the sun would increase their distance, if they were allowed to fall toward the sun, then the sun's action, by endeavoring to separate them, diminishes their gravity to each other; whenever the sun's action would diminish the distance, then it increases their mutual gravitation. Hence, in the conjunction and opposition, that is, *in the syzygies*, their gravity toward each other is diminished by the action of the sun, while in the quadratures it is increased. But it must be remembered that it is not the total action of the sun on them that disturbs their motions, but only that part of it which tends at one time to separate them, and at another time to bring them nearer together. The other and far greater part, has no other effect than to retain them in their annual course around the sun.

227. Suppose the moon setting out from the quarter that precedes the conjunction with a velocity that would make her describe an exact circle round the earth, if the sun's action had no effect on her: since her gravity is increased by that action, she must descend toward the earth and move within that circle. Her orbit then would be more curved than it otherwise would have been; because the addition to her gravity will make her fall further at the end of an arc below the tangent drawn at the other end of it. Her motion will be thus accelerated, and it will continue to be accelerated until she arrives at the ensuing conjunction, because the direction of the sun's action upon her, during that time, makes an acute angle with the direction of her motion. (See Fig. 41.) At the conjunction, her gravity toward the earth being diminished by the action of the sun, her orbit will then be less curved, and she will be carried further from the earth as she moves to the next quarter; and because the action of the sun makes there an obtuse angle with the direction of her motion, she will be retarded in the same degree as she was accelerated before.

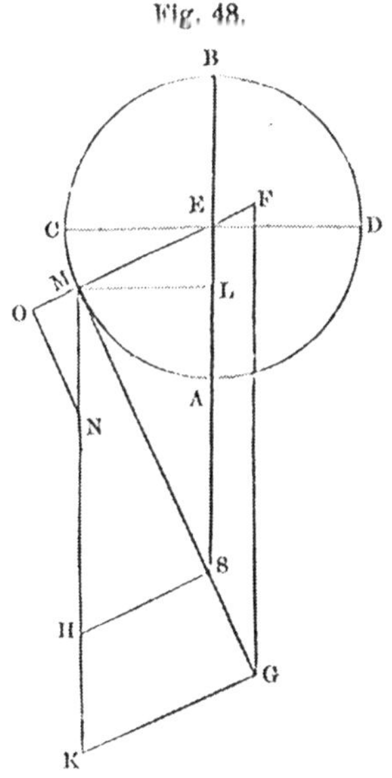

228. After this *general* explanation of the mode in which the sun acts as a disturbing force on the motions of the moon, the learner will be prepared to understand the mathematical development of the same doctrine.

Therefore, let ADBC (Fig. 48) be the orbit, nearly circular, in which the moon M revolves in the direction CADB, round the earth E. Let S be the sun, and let SE, the radius of the earth's orbit, be taken to represent the force with which the earth gravitates to the sun. Then (Art. 181) $\frac{1}{SE^2} : \frac{1}{SM^2} :: SE : \frac{SE^3}{SM^2} =$ the force by which the sun draws the moon in the direction MS. Take MG =

$\frac{SE^3}{SM^2}$, and let the parallelogram KF be described, having MG for its diagonal, and having its sides parallel to EM and ES. The force MG may be resolved into two, MF and MK, of which MF, directed toward E, the center of the earth, increases the gravity of the moon to the earth, and does not hinder the areas described by the radius vector from being proportional to the times. The other force MK draws the moon in the direction of the line joining the centers of the sun and earth. It is, however, only the excess of this force above the force represented by SE, or that which draws the earth to the sun, which disturbs the relative position of the moon and earth. This is evident, for if KM were just equal to ES, no disturbance of the moon, relative to the earth, could arise from it. If, then, ES be taken from MK, the difference HK is the whole force in the direction parallel to SE, by which the sun disturbs the relative position of the moon and earth. Now, if in MK, MN be taken equal to HK, and if NO be drawn perpendicular to the radius vector EM produced, the force MN may be resolved into two, MO and ON, the first lessening the gravity of the moon to the earth; and the second, being parallel to the tangent of the moon's orbit in M, accelerates the moon's motion from C to A, and retards it from A to D, and so alternately in the other two quadrants. Thus the whole solar force directed to the center of the earth, is composed of the two parts MF and MO, which are sometimes opposed to one another, but which never affect the uniform description of the areas about E. Near the quadratures the force MO vanishes, and the force MF, which increases the gravity of the moon to the earth, coincides with CE or DE. As the moon approaches the conjunction at A, the force MO prevails over MF, and lessens the gravity of the moon to the earth. In the opposite point of the orbit, when the moon is in opposition at B, the force with which the sun draws the moon is less than that with which the sun draws the earth, so that the effect of the solar force is to separate the moon and earth, or to increase their distance; that is, it is the same as if, conceiving the earth not to be acted on, the sun's force drew the moon in the direction from E to B. This force is negative, therefore, in respect to the force at A, and the effect in both cases is to draw the moon from the earth in a direction perpendicular to the line of the quadratures. Hence, the general result is, that by the disturbing force of the sun, *the gravity to the earth is increased at the quadratures, and diminished at the syzygies.* It is found by calculation that the average amount of this disturbing force is $\frac{1}{358}$ of the moon's gravity to the earth.*

229. With these general principles in view, we may now proceed to investigate the figure of the moon's orbit, and the irregularities to which the motions of this body are subject.

230. *The figure of the moon's orbit is an ellipse, having the earth in one of the foci.*

The elliptical figure of the moon's orbit, is revealed to us by observations on her changes in apparent diameter, and in her horizontal parallax. First, we may measure from day to day the apparent diameter of the moon. Its variations being inversely as the distances (Art. 163), they give us at once the *relative* distance of each point of observation from the focus. Secondly, the variations on the moon's horizontal parallax, which also are inversely as the distances (Art. 82), lead to the same results. Observations on the angular velocities, combined with the changes in the lengths of the radius vector, afford the means of laying down a plot of the lunar orbit, as in the case of the sun, represented in figure 32. The orbit is shown to be nearly an ellipse, because it is found to have the properties of an ellipse.

The moon's greatest and least apparent diameters are respectively 33′.518 and 29′.365, while her corresponding changes of parallax are 61′.4 and 53′.8. The two ratios ought to be equal, and we shall find such to be the fact very nearly, as expressed by the foregoing numbers; for,

$$61.4 : 53.8 :: 33.518 : 29.369.$$

The greatest and least distances of the moon from the earth, derived from the parallaxes, are 63.8419 and 55.9164, or nearly 64 and 56, the radius of the earth being taken for unity. Hence, taking the arithmetical mean, which is 59.879, we find

* Playfair.

that the mean distance of the moon from the earth is very nearly 60 times the radius of the earth. The point in the moon's orbit nearest the earth, is called her *perigee;* the point furthest from the earth, her *apogee*.

The greatest and least apparent diameters of the *sun* are respectively 32.583, and 31.517, which numbers express also the ratio of the greatest and least distances of the earth from the sun. By comparing this ratio with that of the distances of the moon, it will be seen that the latter vary much more than the former, and consequently that the lunar orbit is much more eccentric than the solar. The eccentricity of the moon's orbit is in fact 0.0548 (the semi-major axis being as usual taken for unity)$=\frac{1}{18}$ of its mean distance from the earth, while that of the earth is only .01685$=\frac{1}{59}$ of its mean distance from the sun.

231. *The moon's nodes constantly shift their positions in the ecliptic from east to west, at the rate of* 19° 35′ *per annum, returning to the same points in* 18.6 *years.*

Suppose the great circle of the ecliptic marked out on the face of the sky in a distinct line, and let us observe, at any given time, the exact point where the moon crosses this line, which we will suppose to be close to a certain star; then, on its next return to that part of the heavens, we shall find that it crosses the ecliptic sensibly to the westward of that star, and so on, further and further to the westward every time it crosses the ecliptic at either node. This fact is expressed by saying that *the nodes retrograde on the ecliptic*, and that the line which joins them, or the line of the nodes, revolves from east to west.

232. This shifting of the moon's nodes implies that the lunar orbit is not a curve returning into itself, but that it more resembles a spiral like the curve represented in figure 49, where *abc* represents the ecliptic, and ABC the lunar orbit, having its nodes at C and E, instead of A and *a*; consequently, the nodes shift backward through the arcs *a*C and AE. The manner in which this effect is produced may be thus explained. That part of the solar force which is parallel to the line joining the

Fig. 49.

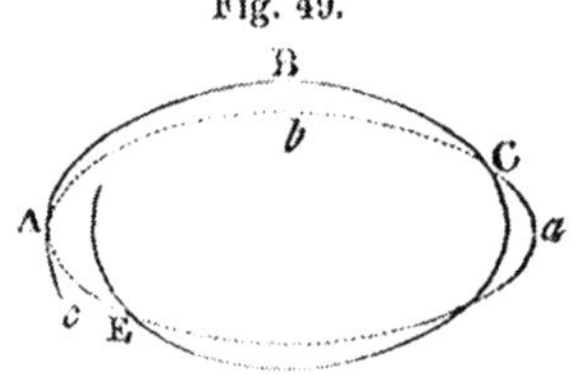

centers of the sun and earth (see Fig. 48), is not in the plane of the moon's orbit (since this is inclined to the ecliptic about 5°), except when the sun itself is in that plane, or when the line of the nodes being produced passes through the sun. In all other cases it is oblique to the plane of the orbit, and may be resolved into two forces, one of which is at right angles to that plane, and is directed toward the ecliptic. This force of course draws the moon continually toward the ecliptic, or produces a continual deflection of the moon from the plane of her own orbit toward that of the earth. Hence the moon meets the plane of the ecliptic sooner than it would have done if that force had not acted. At every half revolution, therefore, the point in which the moon meets the ecliptic shifts in a direction contrary to that of the moon's motion, or contrary to the order of the signs. If the earth and sun were at rest, the effect of the deflecting force just described would be to produce a retrograde motion of the line of the nodes till that line was brought to pass through the sun, and, of consequence, the plan of the moon's orbit to do the same, after which they wo both remain in their position, there being no longer any fo tending to produce change in either. But the motion of the earth carries the line of the nodes out of this position, and produces, by that means, its continual retrogradation. The same force produces a small variation in the inclination of the moon's orbit, giving it an alternate increase and decrease within very narrow limits.* These points will be easily understood by the aid of a diagram. Therefore, let MN (Fig. 50) be the ecliptic, ANB the orbit of the moon, the moon being in L, and N its descending node. Let the disturbing force of the sun which tends to bring it down to the ecliptic be represented by L*b*, and its velocity in its orbit by L*a*. Under the action of these two forces, the moon will describe the diagonal L*c* of the parallelogram *ba*, and its orbit will be changed from AN to LN′; the node N passes to N′; and the exterior angle at N′ of the triangle LNN′ being greater than the interior and opposite angle at N, the inclination of the orbit is increased at the node. After the moon has passed the ecliptic to the south side to *l*, the disturbing force *ld* produces a new change of the orbit N′*lc*

* Playfair.

Fig. 50.

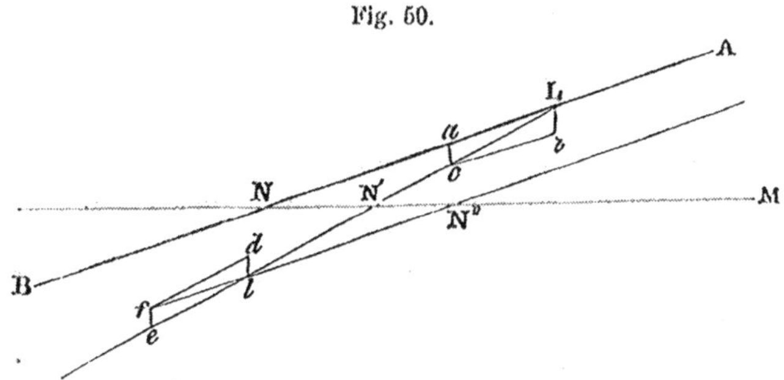

to N''lf, and the inclination is diminished as at N''. In general, while the moon is receding from one of the nodes, its inclination is diminishing; while it is approaching a node, the inclination is increasing.*

233. The period occupied by the sun in passing from one of the moon's nodes until it comes round to the same node again, is called *the synodical revolution of the node.* This period is shorter than the sidereal year, being only about 346½ days. For since the node shifts its place to the westward 19° 35′ per annum, the sun, in his annual revolution, comes to it so much before he completes his entire circuit; and since the sun moves about a degree a day, the synodical revolution of the node is 365—19=346, or more exactly, 346.619851. The time from one new moon, or from one full moon, to another, is 29.5305887 days. Now 19 synodical revolutions of the nodes contain very nearly 223 of these periods.

For 346.619851×19=6585.78,
And 29.5305887×223=6585.32.

Hence, if the sun and moon were to leave the moon's node together, after the sun had been round to the same node 19 times, the moon would have performed very nearly 223 synodical revolutions, and would, therefore, at the end of this period meet at the same node, to repeat the same circuit. And since eclipses of the sun and moon depend upon the relative position of the sun, the moon, and node, these phenomena are repeated in nearly the same order, in each of those periods. Hence,

* Francœur, Uranog., p. 158; Robison's Phys. Astronomy. Art. 264.

this period, consisting of about 18 years and 10 days, under the name of the *Saros*, was used by the Chaldeans and other ancient nations in predicting eclipses.

234. The *Metonic Cycle* is not the same with the Saros, but consists of 19 tropical years. During this period the moon makes very nearly 235 synodical revolutions, and hence the *new and full moons*, if reckoned by periods of 19 years, recur at the same dates. If, for example, a new moon fell on the fiftieth day of one cycle, it would also fall on the fiftieth day of each succeeding cycle; and, since the regulation of games, feasts, and fasts, has been made very extensively according to new or full moons, hence this lunar cycle has been much used both in ancient and modern times. The Athenians adopted it, 433 years before the Christian era, for the regulation of their calendar, and had it inscribed in letters of gold on the walls of the temple of Minerva. Hence the term *Golden Number*, which denotes the year of the lunar cycle.

235. *The line of the apsides of the moon's orbit revolves from west to east through her whole orbit in about nine years.*

If, in any revolution of the moon, we should accurately mark the place in the heavens where the moon comes to its perigee (Art. 230), we should find that at the next revolution it would come to its perigee at a point a little further eastward than before, and so on at every revolution, until, after 9 years, it would come to its perigee at nearly the same point as at first. This fact is expressed by saying that the perigee, and of course the apogee, revolves, and that the line which joins these two points, or the line of the apsides, also revolves.

The place of the perigee may be found by observing when the moon has the greatest apparent diameter. But as the magnitude of the moon varies slowly at this point, a better method of ascertaining the position of the apsides, is to take two points in the orbit where the variations in apparent diameter are most rapid, and to find where they are equal on opposite sides of the orbit. The middle point between the two will give the place of the perigee.

The angular distance of the moon from her perigee in any part of her revolution, is called the *Moon's Anomaly*.

236. The change of place in the apsides of the moon's orbit, like the shifting of the nodes, is caused by the disturbing influence of the sun. If when the moon sets out from its perigee, it were urged by no other force than that of projection, combined with its gravitation toward the earth, it would describe a symmetrical curve (Art. 187), coming to its apogee at the distance of 180°. But as the mean disturbing force in the direction of the radius vector tends, on the whole, to diminish the gravitation of the moon to the earth, the portion of the path described in an instant will be less deflected from her tangent, or less curved, than if this force did not exist. Hence the path of the moon will not intersect the radius vector at right angles, that is, she will not arrive at her apogee until after passing more than 180° from her perigee, by which means these points will constantly shift their positions from west to east.* The motion of the apsides is found to be 3° 1′ 20″ for every sidereal revolution of the moon.

237. On account of the greater eccentricity of the moon's orbit above that of the sun, the *Equation of the Center*, or that correction which is applied to the moon's mean anomaly to find her true anomaly (Art. 200), is much greater than that of the sun, being when greatest more than six degrees (6° 17′ 12″.7), while that of the sun is less than two degrees (1° 55′ 26″.8).

The irregularities in the motions of the moon may be compared to those of the magnetic needle. As a *first approximation*, we say that the needle places itself in a north and south line. On closer examination, however, we find that, at different places, it deviates more or less from this line, and we introduce the first great correction under the name of the *declination* of the needle. But observation shows us that the declination alternately increases and diminishes every day, and therefore we apply to the declination itself a second correction for the *diurnal* variation. Finally, we ascertain, from long-continued observations, that the diurnal variation is affected by the change of seasons, being greater in summer than in winter, and hence we apply to the diurnal variation a third correction for the *annual* variation.

---

* Playfair.

In like manner, we shall find the greater inequalities of the moon's motions are themselves subject to subordinate inequalities which give rise to smaller equations, and these to smaller still, to the last degree of refinement.

238. Next to the equation of the center, the greatest correction to be applied to the moon's longitude, is that which belongs to the *Evection*. The evection is *a change of form in the lunar orbit*, by which its eccentricity is sometimes increased, and sometimes diminished. It depends on the position of the line of the apsides with respect to the line of the syzygies.

This irregularity, and its connection with the place of the perigee with respect to the place of conjunction or opposition, was known *as a fact* to the ancient astronomers, Hipparchus and Ptolemy; but its *cause* was first explained by Newton in conformity with the law of universal gravitation. It was found, by observation, that the equation of the center itself was subject to a periodical variation, being greater than its mean, and greatest of all when the conjunction or opposition takes place at the perigee or apogee, and least of all when the conjunction or opposition takes place at a point half way between the perigee and apogee; or, in the more common language of astronomers, the equation of the center is increased when the line of the apsides is in syzygy, and diminished when that line is in quadrature. If, for example, when the line of the apsides is in syzygy, we compute the moon's place by deducting the equation of the center from the mean anomaly (see Art. 200), seven days after conjunction, the computed longitude will be greater than that determined by actual observation, by about 80 minutes; but if the same estimate is made when the line of the apsides is in quadrature, the computed longitude will be less than the observed, by the same quantity. These results plainly show a connection between the velocity of the moon's motions and the position of the line of the apsides with respect to the line of the syzygies.

239. Now any cause which, at the perigee, should have the effect to increase the moon's gravitation toward the earth beyond its mean, and, at the apogee, to diminish the moon's gravitation toward the earth, would augment the difference between the gravitation at the perigee and at the apogee, and consequently increase the eccentricity of the orbit. Again, any cause which at the perigee should have the effect to lessen the moon's gravitation toward the earth, and at the apogee to increase it, would lessen the difference between the two, and consequently diminish the eccentricity of the orbit, or bring it nearer to a circle. Let us see if the disturbing force of the sun produces these effects. The sun's disturbing force, as we have seen in article 228, admits of two resolutions, one in the direction of the radius vector (OM, Fig. 48), the other (ON) in the direction of a tangent to the orbit. First, let AB be the line of the apsides in syzygy, A being the place of the perigee. The sun's disturbing force OM is greatest in the direction of the line of the syzygies; yet depending as it does on the *unequal* action of the sun upon the earth and the moon, and being greater as their distance from each other is greater, it is at a minimum when acting at the perigee, and at a maximum when acting at the apogee. Hence its effect is to draw away the moon from the earth less than usual at the perigee, and of course to make her gravitation toward the earth greater than usual; while at the apogee its effect is to diminish the tendency of the moon to the earth more than usual, and thus to increase the disproportion between the two distances of the moon from the focus at these two points, and of course to increase the eccentricity of the orbit. The moon, therefore, if moving toward the perigee, is brought to the line of the apsides in a point between its mean place and the earth; or if moving toward the apogee, she reaches the line of the apsides in a point more remote from the earth than its mean place.

Secondly, let CD be the line of apsides in quadrature. The effect of the sun's action is to increase the moon's tendency to the earth, when in quadrature. If, however, the moon is then at perigee, such increase will be *less* than usual, because the sun's action is less oblique; and if at apogee, it will be *more* than usual, because more oblique. Hence its effect will be to lessen the disproportion between the two distances of the moon from the focus at these two points; and of course to diminish the eccentricity of the orbit. The moon, therefore, if moving toward the perigee, meets the line of the apsides in a point more remote from the earth than the mean place of the perigee; and if moving toward the apogee, in a point between the earth and the mean place of the apogee.*

240. A third inequality in the lunar motions, is the *Variation.* By comparing the moon's place as computed from her mean motion corrected for the equation of the center and for evection, with her place as determined by observation, Tycho Brahe discovered that the computed and observed places did not always agree. They agreed only in the syzygies and quadratures, and disagreed most at a point half way between these, that is, at the *octants.* Here, at the maximum, it amounted to more than half a degree (35′ 41″.6). It appeared evident from examining the daily observations while the moon is performing her revolution around the earth, that this inequality is connected with the angular distance of the moon from the sun, and in every part of the orbit could be correctly expressed by multiplying the maximum value, as given above, into the sine of twice the angular distance between the sun and the moon. It is, therefore, 0 at the conjunctions and quadratures, and greatest at the octants. Tycho Brahe knew the *fact:* Newton investigated the *cause.*

It appears by article 228, that the sun's disturbing force can be resolved into two parts: one in the direction of the radius vector, the other at right angles to it in the direction of a tangent to the moon's orbit. The former, as already explained, produces the Evection; the latter produces the Variation. This latter force will accelerate the moon's velocity, in every point of the quadrant which the moon describes in moving from quadrature to conjunction, or from C to A (Fig. 48), but at an unequal rate, the acceleration being greatest at the octant, and nothing at the quadrature and the conjunction; and when the moon is past conjunction, the tangential force will change its direction and retard the moon's motion. All these points will be understood by inspection of figure 48.

241. A fourth lunar inequality is the *Annual Equation.* This depends on the distance of the earth (and of course the moon) from the sun. Since the disturbing influence of the sun

* Woodhouse's Ast., p. 680.

has a greater effect in proportion as the sun is nearer,* consequently all the inequalities depending on this influence must vary at different seasons of the year. Hence, the amount of this effect due to any particular time of the year is called the Annual Equation.

242. The foregoing are the largest of the inequalities of the moon's motions, and may serve as *specimens* of the corrections that are to be applied to the mean place of the moon in order to find her true place. These were first discovered by actual observation; but a far greater number, though most of them are exceedingly minute, have been made known by the investigations of Physical Astronomy, in following out all the consequences of universal gravitation. In the best tables, about 30 equations are applied to the mean motions of the moon. That is, we first compute the place of the moon on the supposition that she moves uniformly in a circle. This gives us her *mean* place. We then proceed, by the aid of the Lunar Tables, to apply the different corrections, such as the equation of the center, evection, variation, the annual equation, and so on, to the number of 28. Numerous as these corrections appear, yet La Place informs us, that the whole number belonging to the moon's longitude is no less than 60; and that to give the tables all the requisite degree of precision, additional investigations will be necessary, as extensive at least as those already made.† The best tables in use in the time of Tycho Brahe, gave the moon's place only by a distant approximation. The tables in use in the time of Newton (Halley's tables), approximated within 7 minutes. Tables at present in use give the moon's place to 5 seconds. These additional degrees of accuracy have been attained only by immense labor, and by the united efforts of Physical Astronomy and the most refined observations.

243. The inequalities of the moon's motions are divided into periodical and secular. *Periodical* inequalities are those which are completed in comparatively short periods, like evec-

* Varying reciprocally as the *cube* of the sun's distance from the earth.

† Syst. du Monde., l. iv., c. 5.

tion and variation; *Secular* inequalities are those which are completed only in very long periods, such as centuries or ages. Hence the corresponding terms *periodical equations*, and *secular equations*. As an example of a secular inequality, we may mention the *acceleration of the moon's mean motion*. It is discovered that the moon actually revolves around the earth in less time now than she did in ancient times. The difference, however, is exceedingly small, being only about 10″ in a century, but increases from century to century as the square of the number of centuries from a given epoch. This remarkable fact was discovered by Dr. Halley.* In a lunar eclipse the moon's longitude differs from that of the sun, at the middle of the eclipse, by exactly 180°; and since the sun's longitude at any given time of the year is known, if we can learn the day and hour when an eclipse occurs, we shall of course know the longitude of the sun and moon. Now in the year 721 before the Christian era, on a specified day and hour, Ptolemy records a lunar eclipse to have happened, and to have been observed by the Chaldeans. The moon's longitude, therefore, for that time, is known; and if we compute the place which the moon would now occupy, had she always maintained her present period of revolution, we shall find that place to be about a degree and a half in advance of her actual place, showing that the period we have used is too short. Moreover, the same conclusion is derived from a comparison of the Chaldean observations with those made by an Arabian astronomer of the tenth century.

This phenomenon at first led astronomers to apprehend that the moon encountered a resisting medium, which, by destroying at every revolution a small portion of her projectile force, would have the effect to bring her nearer and nearer to the earth, and thus to augment her velocity. But in 1786, La Place demonstrated that this acceleration is one of the legitimate effects of the sun's disturbing force, and is so connected with changes in the eccentricity of the earth's orbit, that the moon will continue to be accelerated while that eccentricity diminishes; but when the eccentricity has reached its minimum (as it will do after many ages), and begins to increase, then

* Astronomer Royal of Great Britain, and cotemporary with Sir Isaac Newton.

the moon's motion will begin to be retarded, and thus her mean motions will oscillate forever about a mean value.

244. The lunar inequalities which have been considered are such only as affect the moon's longitude; but the sun's disturbing force also causes inequalities in the moon's *latitude* and *parallax*. Those of latitude alone, require no less than twelve equations. Since the moon revolves in an orbit inclined to the ecliptic, it is easy to see that the oblique action of the sun must admit of a resolution into two forces, one of which being perpendicular to the moon's orbit, must effect changes in her latitude. Since, also, several of the inequalities already noticed, involve changes in the length of the radius vector, it is obvious that the moon's parallax must be subject to corresponding perturbations.

CHAPTER VII.

ECLIPSES.

245. An *eclipse of the moon* happens when the moon, in its revolution about the earth, falls into the earth's shadow. An *eclipse of the sun* happens when the moon, coming between the earth and the sun, covers either a part or the whole of the solar disk. An eclipse of the sun can occur only at the time of conjunction, or new moon; and an eclipse of the moon only at the time of opposition, or full moon. Were the moon's orbit in the same plane with that of the earth, or did it coincide with the ecliptic, then an eclipse of the sun would take place at every conjunction, and an eclipse of the moon at every opposition; for as the sun and earth both lie in the ecliptic, the shadow of the earth must also extend in the same plane, being, of course, always directly opposite to the sun; and since, as we shall soon see, the length of this shadow is much greater than the distance of the moon from the earth, the moon, if it revolved in the plane of the ecliptic, must pass through the shadow at every full moon. For similar reasons, the moon would occasion an eclipse of the sun, partial or total, in some portions of the earth at every new moon. But the lunar orbit is inclined to the ecliptic about 5°, so that the center of the moon, when she is furthest from her node, is 5° from the axis of the earth's shadow (which is always in the ecliptic); and, as we shall show presently, the greatest distance to which the shadow extends on each side of the ecliptic, that is, the greatest semi-diameter of the shadow, where the moon passes through it, is only about $\frac{3}{4}$ of a degree, while the semi-diameter of the moon's disk is only about $\frac{1}{4}$ of a degree; hence the two semi-diameters, namely, that of the moon and the earth's shadow, cannot overlap one another, unless, at the time of new or full moon, the sun is at or very near the moon's node. In the course of the sun's apparent revolution around the earth once a year, he is successively in every part of the ecliptic; consequently, the conjunctions and oppositions of the sun and moon may occur at any part of the ecliptic, either when the sun is at the moon's node (or when he is passing that point of the celestial vault on which the moon's node is projected as seen from the earth), or they may occur when the sun is 90° from the moon's node, where the lunar and solar orbits are at the greatest distance from each other; or, finally, they may occur at any intermediate point. Now the sun, in his annual revolution, passes each of the moon's nodes on opposite sides of the ecliptic, and of course at opposite seasons of the year; so that, for any given year, the eclipses commonly happen in two opposite months, as January and July, February and August, May and November. These are called *Node Months*, and become earlier each year, because the nodes retrograde.

246. If the sun were of the same size with the earth, the shadow of the earth would be cylindrical and infinite in length, since the tangents drawn from the sun to the earth (which form the boundaries of the shadow) would be parallel to each other; but as the sun is a vastly larger body than the earth, the tangents converge and meet in a point at some distance behind the earth, forming a cone, of which the earth is the base, and whose vertex (and of course its axis) lies in the ecliptic. A little reflection will also show us that the form and dimensions of the shadow must be affected by several circumstances; that the shadow must be of the greatest length and breadth when the sun is furthest from the earth; that its figure will be slightly modified by the spheroidal figure of the earth; and that the moon, being, at the time of its opposition, sometimes nearer to the earth, and sometimes further from it, will accordingly traverse it at points where its breadth varies more or less.

247. *The semi-angle of the cone of the earth's shadow is equal to the sun's apparent semi-diameter, minus his horizontal parallax.*

Let AS (Fig. 51) be the semi-diameter of the sun, BE that of the earth, and EC the axis of the earth's shadow. Then the semi-angle of the cone of the earth's shadow

Fig. 51.

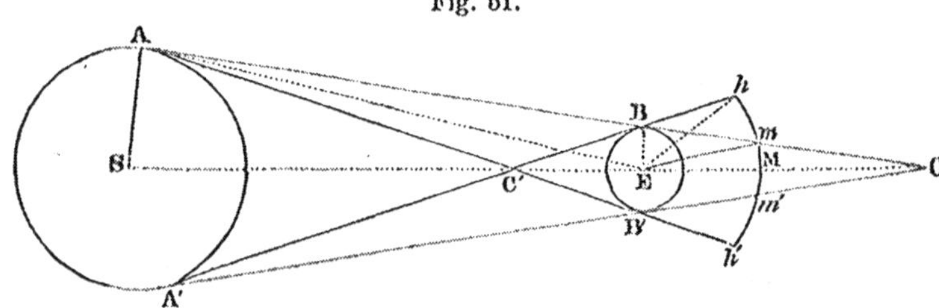

ECB = AES − EAB, of which AES is the sun's semi-diameter, and EAB his horizontal parallax; and as both these quantities are known, hence the angle at the vertex of the shadow becomes known. Putting δ for the sun's semi-diameter, and p for his horizontal parallax, we have the semi-angle of the earth's shadow ECB $= \delta - p$.

248. *At the mean distance of the earth from the sun, the length of the earth's shadow is about* 860,000 *miles, or more than three times the distance of the moon from the earth.*

In the right-angled triangle ECB, right angled at B, the angle ECB being known, and the side EB, we can find the side EC. For $\sin(\delta - p) : \text{EB} :: \text{R} : \text{EC} = \dfrac{\text{EB}}{\sin(\delta - p)}$. This value will vary with the sun's semi-diameter, being greater as that is less. Its mean value being 16′ 1″.5, and the sun's horizon tal parallax being 8″.6, $\delta - p = 15' \ 52''.9$, and EB = 3956.2 Hence,

Sin 15′ 53″ : Rad : : 3956.2 : 856,275.

Since the distance of the moon from the earth is 238,545 miles, the shadow extends about 3.6 times as far as the moon, and consequently the moon passes the shadow toward its broadest part, where its breadth is much more than sufficient to cover the moon's disk.

249. *The average breadth of the earth's shadow, where it eclipses the moon, is almost three times the moon's diameter.*

Let mm' (Fig. 51) represent a section of the earth's shadow where the moon passes through it, M being the center of the circular section. Then the angle MEm will be the angular breadth of half the shadow. But,

MEm = BmE − BCE; that is, since BmE is the moon's horizontal parallax (Art. 82), and BCE equals the sun's semi-diameter minus his horizontal parallax ($\delta - p$), therefore, putting P for the moon's horizontal parallax, we have

MEm = P − ($\delta - p$) = P + p − δ; that is, since P = 57′ 1″, and $\delta - p = 15' \ 52''.9$, therefore 57′ 1″ − 15′ 52″.9 = 41′ 8″.1, which is nearly three times 15′ 33″, the semi-diameter of the moon. Thus it is seen how, by the aid of geometry, we learn to estimate various particulars respecting the earth's shadow, by means of simple data derived from observation.

250. The distance from the node to the center of the earth's shadow, when so situated that the moon would merely touch it at opposition, is called the *lunar ecliptic limit;* and the *solar ecliptic limit* is the distance from the node to the center of the sun's disk, when the moon apparently touches it in passing the conjunction. All eclipses of the moon and sun occur within these limits respectively.

251. *The Lunar Ecliptic Limit is nearly 12 degrees.*

Let CN (Fig. 52) be the sun's path, MN the moon's, and N the node. Let Ca be the semi-diameter of the earth's shadow, and Ma the semi-diameter of the moon. Since Ca and Ma are known quantities, their sum CM is also known. The angle at N is known, being the inclination of the lunar orbit to the ecliptic. Hence, in the spherical triangle MCN, right angled at M,* by Napier's theorem (Art. 132, *Note*),

$$\text{Rad} \times \sin \text{CM} = \sin \text{CN} \times \sin \text{MNC}.$$

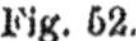

Fig. 52.

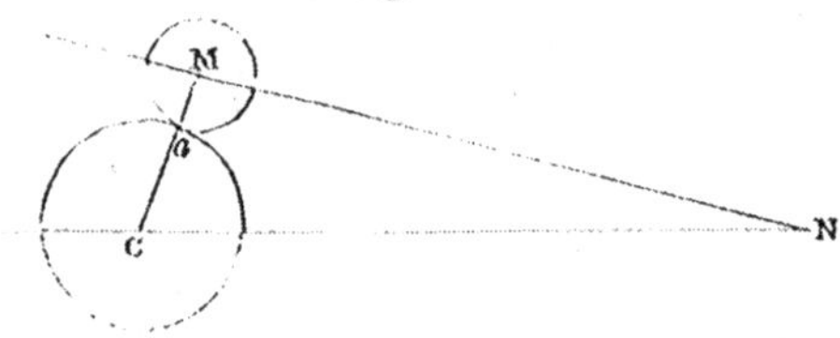

The greatest apparent semi-diameter of the earth's shadow where the moon crosses it, computed by article 249, is 45′ 52″, and the moon's greatest apparent semi-diameter is 16′ 45″.5, which together give MC equal to 62′ 37″.5. Taking the inclination of the moon's orbit, or the angle MNC (what it generally is in these circumstances) at 5° 17′, and we have $\text{Rad} \times \sin 62' \, 37''.5 = \sin \text{CN} \times \sin 5° \, 17'$, or $\sin \text{CN} = \dfrac{\text{Rad} \times \sin 62' \, 37''.5}{\sin 5° \, 17'}$, and $\text{CN} = 11° \, 25' \, 40''$.† This is the greatest distance of the moon from her node at which an eclipse of the moon can take place. By varying the value of CM, corresponding to variations in the distances of the sun and moon from the earth, it is found that if NC is less than 9°, there *must* be an eclipse; but between this and the limit, the case is doubtful.

When the moon's disk only comes in contact with the earth's shadow, as in figure 52, the phenomenon is called an *appulse;* when only a part of the disk enters the shadow, the eclipse is said to be *partial*, and *total* if the whole of the disk enters the shadow. The eclipse is called *central* when the moon's center coincides with the axis of the shadow, which happens when the moon at the moment of opposition is exactly at her node.

252. Before the moon enters the earth's shadow, the earth begins to intercept from it portions of the sun's light, gradually increasing until the moon reaches the shadow. This partial light is called the moon's *Penumbra.* Its limits are ascer-

* The line CM is to be regarded as the *projection* of the line which connects the centers of the moon and section of the earth's shadow, as seen from the earth.

† Woodhouse's Astronomy, p. 718.

tained by drawing the tangents AC'B' and A'C'B. (Fig. 51.) Throughout the space included between these tangents, more or less of the sun's light is intercepted from the moon by the interposition of the earth; for it is evident, that as the moon moves toward the shadow, she would gradually lose the view of the sun, until, on entering the shadow, the sun would be entirely hidden from her.

253. *The semi-angle of the Penumbra equals the sun's semi-diameter and horizontal parallax, or* $\delta+p$.

The angle hC'M (Fig. 51) = AC'S = AES+B'AE. But AES is the sun's semi-diameter, and B'AE is the sun's horizontal parallax, both of which quantities are known.

254. *The semi-angle of a section of the Penumbra, where the moon crosses it, equals the moon's horizontal parallax, plus the sun's, plus the sun's semi-diameter.*

The angle hEM (Fig. 51) = EhC'+EC'h. But EhC'=P, the moon's horizontal parallax, and EC'h=$\delta+p$ (Art. 253), ∴ hEM=P+p+δ, all which are likewise known quantities.

Hence, by means of these few elements, which are known from observation, we ascend to a complete knowledge of all the particulars necessary to be known respecting the moon's penumbra.

255. In the preceding investigations, we have supposed that the cone of the earth's shadow is formed by lines drawn from the sun, and touching the earth's surface. But the apparent diameter of the shadow is found by observation to be somewhat greater than would result from this hypothesis. The fact is accounted for by supposing that a portion of the solar rays which graze the earth's surface are absorbed and extinguished by the lower strata of the atmosphere. This amounts to the same thing as though the earth were larger than it is, in which case the moon's horizontal parallax would be increased; and accordingly, in order that theory and observation may coincide, it is found necessary to increase the parallax by $\frac{1}{60}$.

256. In a total eclipse of the moon, its disk is still visible, shining with a dull red light. This is due to the earth's atmosphere, which acts as a convex lens, and converges the light into the shadow. The lower rays, if they could escape, would be bent twice 34′ (Art. 89), and reach the axis thousands of miles this side of the moon. As it is, only a little light emerges, which is sufficiently bent to fall on the moon when centrally eclipsed. An observer at the moon, in witnessing a *solar* eclipse, would see the sun expanded into a dim narrow ring, having nearly four times its usual diameter.

257. In *calculating an eclipse of the moon*, we first learn from the tables in what month the sun, at the time of full moon in that month, is near the moon's node, or within the lunar ecliptic limit. This it must evidently be easy to determine, since the tables enable us to find both the longitudes of the nodes, and the longitudes of the sun and moon, for every day of the year. Consequently, we can find when the sun has nearly the same longitude as one of the nodes, and also the precise moment when the longitude of the moon is 180° from that of the sun, for this is the time of opposition, from which may be derived the time of the middle of the eclipse. Having the time of the middle of the eclipse, and the breadth of the shadow (Art. 249), and knowing, from the tables, the rate at which the moon moves per hour faster than the shadow, we can find how long it will take her to traverse half the breadth of the shadow; and this time subtracted from the time of the middle of the eclipse, will give the beginning, and added to the time of the middle will give the end of the eclipse. Or if instead of the breadth of the shadow, we employ the breadth of the penumbra (Art. 253), we may find, in the same manner, when the moon enters and when she leaves the penumbra. We see, therefore, how, by having a few things known by observation, such as the sun and moon's semi-diameters, and their horizontal parallaxes, we rise, by the aid of trigonometry, to the knowledge of various particulars respecting the length and breadth of the shadow and of the penumbra. These being known, we next have recourse to the tables which contain all the necessary particulars respecting the motions of the sun and moon, together with equations or corrections, to be applied for all their irregularities. Hence it is comparatively an easy task to calculate with great accuracy an eclipse of the moon.

258. Let us then see how we may find the *exact time* of the beginning, end, duration, and magnitude, of a lunar eclipse.

Let NG (Fig. 53) be the ecliptic, and *Nag* the moon's orbit, the sun being in A* when the moon is in opposition at *a*; let N be the ascending node, and A*a* the moon's latitude at the

Fig. 53.

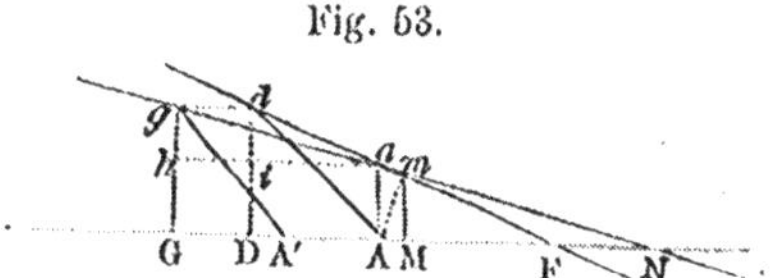

instant of opposition. An hour afterward the sun will have passed to A', and the moon to *g*, when the difference of longitude of the two bodies will be GA'. Then *gh* is the moon's hourly motion in latitude, and *ah* her hourly motion in longitude. As the character and form of the eclipse will depend solely upon the distances between the centers of the sun and moon, that is, upon the line *g*A', instead of considering the two bodies as both in motion, we may suppose the sun at rest in A, and the moon as advancing with a motion equal to the difference between its rate and that of the sun, a supposition that will simplify the calculation. Therefore, draw *gd* parallel and equal to A'A, join *d*A, and this line being equal to *g*A', the two bodies will be in the same relative situation as if the sun were at A' and the moon at *g*. Join *da* and produce the line *da* both ways, cutting the ecliptic in F; then *da*F will be the moon's *Relative Orbit.* Hence $ai = ah - AA'$ = the difference of the hourly motions of the sun and moon, that is, the moon's *relative motion* in longitude, and *di* = the moon's hourly motion in latitude.

Draw CD (Fig. 54) to represent the ecliptic, and let A be the place of the sun. As the tables give the computation of the moon's latitude at every instant, consequently, we may take from them the latitude corresponding to the instant of opposition, and to one hour later; and we may take also the sun's and moon's hourly motions in longitude. Take AD, AB, each

* It will be remarked that the point A really represents the center of the earth's shadow; but as the *real* motions of the shadow are the same with the *assumed* motions of the sun, the latter are used in conformity with the language of the tables.

equal to the relative motion, and Aa = the latitude in opposition, Dd = the latitude one hour afterward; join da and produce the line da both ways, and it will represent the moon's relative orbit. Draw Bb at right angles to CD, and it will be the lati-

Fig. 54.

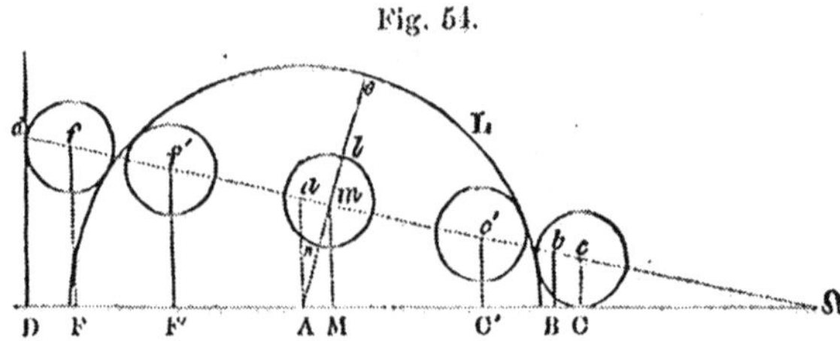

tude an hour before opposition. At the time of the eclipse, the apparent distance of the center of the shadow from the moon is very small; consequently, CD, cd, Dd, &c., may be regarded as straight lines. During the short interval between the beginning and end of an eclipse, the motion of the sun, and consequently that of the center of the shadow, may likewise be regarded as uniform.

259. The various particulars that enter into the calculation of an eclipse are called its *Elements;* and as our object is here merely *to explain the method* of calculating an eclipse of the moon (referring to the *Supplement* for the actual computation), we may take the elements at their mean value. Thus, we will consider cd as inclined to CD 5° 9′, the moon's horizontal parallax as 58′, its semi-diameter as 16′, and that of the earth's shadow as 42′. The line Am, perpendicular to cd, gives the point m for the place of the moon at the middle of the eclipse, for this line bisects the chord, which represents the path of the moon through the shadow; and mM, perpendicular to CD, gives AM for the time of the middle of the eclipse before opposition, the number of minutes before opposition being the same part of an hour that AM is of AB.* From the center A, with a radius equal to that of the earth's shadow (42′), describe the semicircle BLF, and it will represent the

* The situation of the moon when at m is called *orbit opposition;* and her situation when at a, *ecliptic opposition.*

projection of the shadow traversed by the moon. With a radius equal to the semi-diameter of the shadow and that of the moon ($=42'+16'=58'$), and with the center A, mark the two points *c* and *f* on the relative orbit, and they will be the places of the center of the moon at the beginning and end of the eclipse. The perpendiculars *c*C, *f*F, give the times AC and AF of the commencement and the end of the eclipse, and CM or MF gives half the duration. From the centers *c* and *f*, with a radius equal to the semi-diameter of the moon (16′), describe circles, and they will each touch the shadow (Euc., 3, 12), indicating the position of the moon at the beginning and end of the eclipse. If the same circle described from *m* is wholly within the shadow, the eclipse will be *total*; if it is only partly within the shadow, the eclipse will be *partial*. With the center A, and radius equal to the semi-diameter of the shadow minus that of the moon ($42'-16'=26'$), mark the two points c', f', which will give the places of the center of the moon, at the beginning and end of total darkness, and MC′, MF′ will give the corresponding times before and after the middle of the eclipse. Their sum will be the duration of total darkness.

260. If the foregoing projection be accurately made from a scale, the required particulars of the eclipse may be ascertained by measuring, on the same scale, the lines which respectively represent them; and we should thus obtain a near approximation to the elements of the eclipse. A more accurate determination of these elements may, however, be obtained by actual calculation. The general principles of the calculation will be readily understood.

First, knowing *ai* (Fig. 53), the moon's relative longitude, and *di*, her latitude, we find the angle *dai*, which is the inclination of the moon's relative orbit. But $dai=a\text{A}m$; and, in the triangle *a*A*m*, we have the angle at A, and the side A*a*, being the moon's latitude at the time of opposition, which is given by the tables. Hence we can find the side A*m*. In the triangle A*m*M (Fig. 54), having the side A*m* and the angle A*m*M ($=a\text{A}m$), we can find AM = the arc of relative longitude described by the moon from the time of the middle of the eclipse to the time of opposition; and knowing the moon's hourly motion in longitude, we can convert AM into time, and this subtracted from the time of opposition gives us the *time of the middle of the eclipse.*

Secondly, joining Ac (Fig. 54), not represented in the figure, we calculate mAc, in the triangle Acm; then mAB (= maA) − mAc = cAC. Hence, in the triangle ACc, we can determine AC, and therefore AC − AM = MC, which, changed into time as before, gives us, when subtracted from the time of the middle of the eclipse, *the time of the beginning of the eclipse*, or, when added to that of the middle, *the time of the end of the eclipse.* The sum of the two equals the *whole duration.*

Thirdly, by a similar method we calculate the value of MC′, which converted into time, and subtracted from the time of the middle of the eclipse, gives the *commencement of total darkness*, or when added gives the *end of total darkness.* Their sum is the *duration of total darkness.*

Fourthly, the *quantity of the eclipse* is determined by supposing the diameter of the moon divided into twelve equal parts called *Digits*, and finding how many such parts lie within the shadow, at the time when the centers of the moon and the shadow are nearest to each other. Even when the moon lies wholly within the shadow, the quantity of the eclipse is still expressed by the number of digits contained in that part of the line which joins the center of the shadow and the center of the moon, which is intercepted between the edge of the shadow and the inner edge of the moon. Thus in figure 54, the number of digits eclipsed, equals $\frac{no}{\frac{1}{12}nl} = \frac{Ao - An}{\frac{1}{12}nl} = \frac{Ao - (Am - nm)}{\frac{1}{12}nl}$, an expression containing only known quantities.

261. The foregoing will serve as an explanation of the *general principles*, on which proceeds the calculation of a lunar eclipse. The actual methods practiced employ many expedients to facilitate the process, and to insure the greatest possible accuracy, the nature of which are explained and exemplified in Mason's Supplement to this work.

262. The leading particulars respecting an ECLIPSE OF THE

SUN, are ascertained very nearly like those of a lunar eclipse. The shadow of the moon travels over a portion of the earth, as the shadow of a small cloud, seen from an eminence in a clear day, rides along over hills and plains. Let us imagine ourselves standing on the moon; then we shall see the earth partially eclipsed by the shadow of the moon, in the same manner as we now see the moon eclipsed by the earth's shadow; and we might proceed to find the length of the shadow, its breadth where it eclipses the earth, the breadth of the penumbra, and its duration and quantity, in the same way as we have ascertained these particulars for an eclipse of the moon.

But, although the general characters of a solar eclipse might be investigated on these principles, so far as respects the earth at large, yet as the appearances of the same eclipse of the sun are very different at different places on the earth's surface, it is necessary to calculate its peculiar aspects for each place separately, a circumstance which makes the calculation of a solar eclipse much more complicated and tedious than of an eclipse of the moon. The moon, when she enters the shadow of the earth, is deprived of the light of the part immersed, and that part appears black alike to all places where the moon is above the horizon. But it is not so with a solar eclipse. We do not see this by the shadow cast on the earth, as we should do if we stood on the moon, but by the interposition of the moon between us and the sun; and his edge may be hidden from one observer while he is in full view of another only a few miles distant. In strictness, the phenomenon should be called an *occultation*, not an *eclipse*, of the sun; the *earth* is eclipsed, or obscured by a shadow cast upon it, while the *sun*, to those within the shadow, is *hidden* by an intervening body.

263. We have compared the motion of the moon's shadow over the surface of the earth to that of a cloud; but its *velocity* is incomparably greater. The mean motion of the moon around the earth is about 33′ per hour; but 33′ of the moon's orbit is 2280 miles, and the shadow moves of course at the same rate, or 2280 miles per hour, traversing the entire disk of the earth in less than four hours. This is the velocity of the shadow when it passes *perpendicularly* over the earth; when the direction of the axis of the shadow is oblique to the earth's surface, the velocity is increased in proportion of radius to the sine of obliquity. Thus the shadows of evening have a far more rapid motion than those of noonday.

Let us endeavor to form a just conception of the manner in which these three bodies—the sun, the earth, and the moon—are situated with respect to each other at the time of a solar eclipse. First, suppose the conjunction to take place at the node. Then the straight line which connects the centers of the sun and the earth, also passes through the center of the moon, and coincides with the axis of its shadow; and, since the earth is bisected by the plane of the ecliptic, the shadow would traverse the earth in the direction of the terrestrial ecliptic, from west to east, passing over the middle regions of the earth. Here the diurnal motion of the earth being in the same direction with the shadow, but with a less velocity, the shadow will appear to move with a speed equal only to the difference between the two. Secondly, suppose the moon is on the north side of the ecliptic at the time of conjunction, and moving toward her descending node, and that the conjunction takes place just within the solar ecliptic limit, say 16° from the node. The shadow will now not fall in the plane of the ecliptic, but a little northward of it, so as just to graze the earth near the pole of the ecliptic. The nearer the conjunction comes to the node, the further the shadow will fall from the pole of the ecliptic toward the equatorial regions. In certain cases, the shadow strikes beyond the pole of the earth; and then its easterly motion being opposite to the diurnal motion of the places which it traverses, consequently its velocity is greatly increased, being equal to the sum of both.

264. After these general considerations, we will now examine more particularly the method of investigating the elements of a solar eclipse.

The *length of the moon's shadow* is the first object of inquiry. The moon, as well as the earth, is at different distances from the sun at different times, and hence the length of her shadow varies, being always greatest when she is furthest from the sun. Also, since her distance from the earth varies, the section of the moon's shadow made by the earth, is greater in proportion as the moon is nearer the earth. The greatest eclipses of the sun, therefore, happen when the sun is in apogee,* and the moon in perigee.

265. *When the moon is at her mean distance from the earth, and from the sun, her shadow nearly reaches the earth's surface.*

Let S (Fig. 55) represent the sun, D the moon, and T the earth. Then, the semi-angle of the cone of the moon's shadow,

Fig.55.

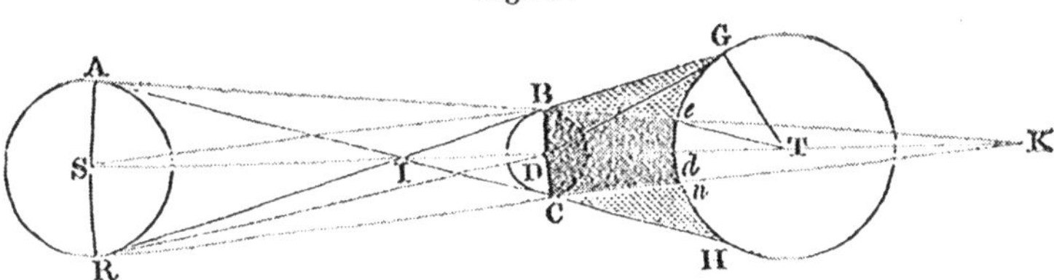

DKR, will, as in the case of the earth (Art. 247), equal SDR–DRK, of which SDR is the sun's apparent semi-diameter, as seen from the moon, and DRK, is the sun's horizontal parallax at the moon. Since, on account of the great distance of the sun compared with that of the moon, the semi-diameter of the sun as seen from the moon must evidently be very nearly the same as when seen from the earth, and since, on account of the minuteness of the moon's semi-diameter when seen from the sun, the sun's horizontal parallax at the moon must be very small, we might, without much error, take the sun's apparent semi-diameter from the earth, as equal to the semi-angle of the cone of the moon's shadow; but, for the sake of greater accuracy, let us estimate the value of the sun's semi-diameter and horizontal parallax at the moon.

Now, SDR : STR : : ST : SD† : : 400 : 399; hence SDR$=\frac{400}{399}$STR$=1.0025$ STR; and the sun's mean semi-diameter STR being 16.025, hence SDR$=1.0025\times16.025=16.065=16'\ 3''.9$.

Again, since parallax is inversely as the distance, the sun's

* The sun is said to be in apogee, when the earth is in aphelion.

† The apparent magnitude of an object being reciprocally as its distance from the eye. See Note, p. 87.

horizontal parallax at the moon, is, since she is nearer to the sun, about $\frac{1}{400}$ greater than at the earth; but on account of her inferior size it is $\frac{7912}{2160}$ less than at the earth. Hence, increasing the sun's horizontal parallax at the earth by the former fraction, and diminishing it by the latter, we have $\frac{400}{399}\times\frac{2160}{7912}\times 9''=2''.5=$the sun's horizontal parallax at the moon. Therefore, the semi-angle of the cone of the moon's shadow, which, as appears above, equals SDR−DRK, equals $16'\ 3''.9-2''.5=16'\ 1''.4$, which so nearly equals the sun's apparent semi-diameter, as seen from the earth, that we may adopt the latter as the value of the semi-angle of the shadow. Hence, sin $16'\ 1''.5$: 1080 (BD) : : Rad : DK=231690. But the mean distance of the moon from the surface of the earth is 238545−3956=234589, which exceeds a little the mean length of the shadow as above.

But when the moon is nearest the earth, her distance from the center of the earth is only 221,148 miles; and when the earth is furthest from the sun, the sun's apparent semi-diameter is only $15'\ 45''.5$. By employing this number in the foregoing estimate, we shall find the length of the shadow 235,630 miles; and 235630−221148=14482, the distance which the moon's shadow may reach beyond the center of the earth.

266. *The diameter of the moon's shadow where it traverses the earth, is, at its maximum, about* 170 *miles.**

In the triangle *e*TK, the angle at K=$15'\ 45''.5$ (Art. 265), the side T*e*=3956, and TK=14482.

Or, 3956 : 14482 : : sin $15'\ 45''.5$: sin $57'\ 41''.5$.

And $57'\ 41''.5+15'\ 45''.5=1°\ 13'\ 27''=d$T*e*, or the arc *de*.

And 2*de*=$2°\ 26'\ 54''$=*en*.

Hence 360 : 2.45 (=$2°\ 26'\ 54''$) : : 24899† : 170 (nearly).

267. *The greatest portion of the earth's surface ever covered by the moon's penumbra, is about* 4393 *miles.*

* This supposes the conjunction to take place at the node, and the shadow to strike the earth perpendicularly to its surface; where it strikes obliquely, the section may be greater than this.

† The equatorial circumference.

The semi-angle of the penumbra $BID=BSD+SBR$, of which BSD the sun's horizontal parallax at the moon$=2''.5$, and SBR the sun's apparent semi-diameter$=16'\ 3''.9$, and hence BID is known. The moon's apparent semi-diameter $BGD=16'\ 45''.5$. Therefore GDT is known, as likewise DT and TG. Hence the angle GTd may be found, and the arc dG and its double GH, which equals the angular breadth of the penumbra It may be converted into miles by stating a proportion as in article 266. On making the calculation it will be found to be 4393 miles.

268. The apparent diameter of the moon is sometimes larger than that of the sun, sometimes smaller, and sometimes exactly equal to it. Suppose an observer placed on the right line which joins the centers of the sun and moon; if the apparent diameter of the moon is greater than that of the sun, the eclipse will be total. If the two diameters are equal, the moon's shadow just reaches the earth, and the sun is hidden but for a moment from the view of spectators situated in the

Fig. 55'.

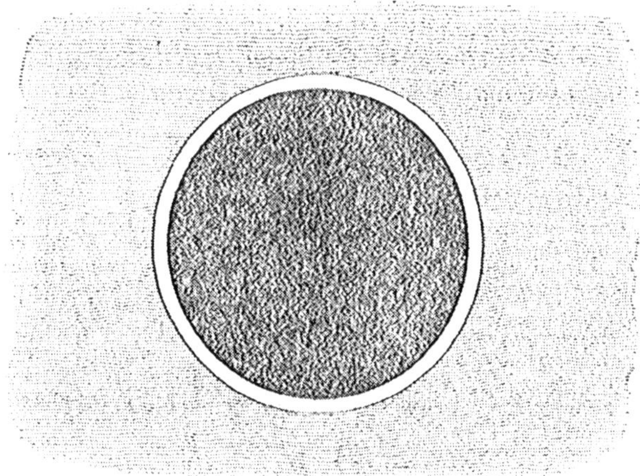

line which the vertex of the shadow describes on the surface of the earth. But if, as happens when the moon comes to her conjunction in that part of her orbit which is toward her apogee, the moon's diameter is less than the sun's, then the observer will see a ring of the sun encircle the moon, constituting an annular eclipse. (Fig. 55'.)

269. Eclipses of the sun are modified by the elevation of the moon above the horizon, since its apparent diameter is augmented as its altitude is increased (Art. 217). This effect, combined with that of parallax, may so increase or diminish the apparent distance between the centers of the sun and moon, that from this cause alone, of two observers at a distance from each other, one might see an eclipse which was not visible to the other. If the horizontal diameter of the moon differs but little from the apparent diameter of the sun, the case might occur where the eclipse would be annular over the places where it was observed morning and evening, but total where it was observed at mid-day.

The earth in its diurnal revolution and the moon's shadow both move from west to east, but the shadow moves faster than the earth; hence the moon overtakes the sun on its western limb and crosses it from west to east. The excess of the apparent diameter of the moon above that of the sun in a total eclipse is so small, that total darkness seldom continues longer than four minutes, and can never continue so long as eight minutes. An annular eclipse may last 12m. 24s.

Since the sun's ecliptic limits are more than 17° and the moon's less than 12°, eclipses of the sun are more frequent than those of the moon. Yet lunar eclipses being visible to every part of the terrestrial hemisphere opposite to the sun, while those of the sun are visible only to the small portion of the hemisphere on which the moon's shadow falls, it happens that for any particular place on the earth, lunar eclipses are more frequently visible than solar. In any year, the number of eclipses of both luminaries can not be less than two nor more than seven: the most usual number is four, and it is very rare to have more than six. The sun does not remain long enough near a node for the moon to be in syzygy, within ecliptic limits, more than three times. Hence, only three eclipses can occur successively while the sun is near a node. As he passes both nodes in the same year, there may therefore be six eclipses. But a seventh may possibly come just within 12 months, reckoned from the first, in consequence of the backward motion of the nodes.

270. It has been observed already, that were the spectator on the moon instead of on the earth, he would see the earth eclipsed by the moon, and the calculation of the eclipse would be very similar to that of a lunar eclipse; but to an observer on the earth the eclipse does not of course begin when the earth first enters the moon's shadow, and it is necessary to determine not only what portion of the earth's surface will be covered by the moon's shadow, but likewise the path described by its center relative to various places on the surface of the earth. This is known when the latitude and longitude of the center of the shadow on the earth is determined for each instant. The latitude and longitude of the moon are found on the supposition that the spectator views it from the center of the earth, whereas his position on the surface changes, in consequence of parallax, both the latitude and longitude, and the amount of these changes must be accurately estimated, before the appearance of the eclipse at any particular place can be fully determined.

The details of the method of calculating a solar eclipse can not be understood in any way so well as by actually performing the process according to a given example. For such details, therefore, the reader is referred to the *Supplement.*

271. In total eclipses there has sometimes been seen a remarkable radiation of light around the moon, while the sun is behind it. This is called a *corona*, and appears to be concentric with the sun's disk, rather than with the moon's. It is by some considered to indicate the existence of an extensive solar atmosphere.

Another interesting phenomenon often attends the moment of concealment and reappearance of the sun's edge at the beginning and end of total darkness, as also the formation and rupture of the ring in annular eclipses. It is the dividing up of the fine thread of the sun's edge into a series of bright beads. Being first noticed by Mr. Francis Baily, they are known by the name of *Baily's Beads.* The appearance is by some attributed to the light of the sun's edge coming through between the mountain summits of the rough outline of the moon's disk. That they are not always seen may arise from the fact that the edge may in some cases be much less serrated by mountains than in others.*

There is another peculiarity for which no satisfactory explanation is yet offered. The moment succeeding the total occultation of the sun, irregular projections start out from the edge here and there, entirely detached from each other, either of a flame or a rose color, sometimes short and wide, at others narrow, 2′ or 3′ long, and often bent at a considerable angle. During the continuance of the eclipse, these change their forms, or disappear, and new ones start up elsewhere. The instant that the thread-like edge of the sun appears, the white corona and the flame-colored protuberances vanish.

A total eclipse of the sun is one of the most sublime and impressive phenomena of nature. The darkness is such that the larger planets and stars appear, and a chill is felt like that of night. Flowers shut up, and animals retire to rest. It is not strange that people of barbarous countries are filled with consternation and fear by the occurrence of a total eclipse.

CHAPTER VIII.

LONGITUDE—TIDES.

272. As eclipses of the sun afford one of the most approved methods of finding the longitudes of places, our attention is naturally turned next toward that subject.

The ancients studied astronomy in order that they might read their destinies in the stars; the moderns, that they may securely navigate the ocean. A large portion of the refined labors of modern astronomy has been directed toward perfecting the astronomical tables, with the view of finding the longitude at sea—an object manifestly worthy of the highest efforts

* Lardner.

of science, considering the vast amount of property and of human life involved in navigation.

273. *The difference of longitude between two places, may be found by any method by which we can ascertain the difference of their local times at the same instant of absolute time.*

As the earth turns on its axis from west to east, any place that lies eastward of another will come sooner under the sun, or will have the sun earlier on the meridian, and consequently, in respect to the hour of the day, will be in advance of the other at the rate of one hour for every 15°, or four minutes of time for each degree. Thus, to a place 15° east of Greenwich, it is 1 o'clock, P. M., when it is noon at Greenwich; and to a place 15° west of that meridian, it is 11 o'clock, A. M. at the same instant. Hence, the difference of time at any two places indicates their difference of longitude.

274. The easiest method of finding the longitude is by means of an accurate timepiece, or *chronometer*. Let us set out from London, with a chronometer accurately adjusted to Greenwich time, and travel eastward to a certain place, where the time is accurately kept, or may be ascertained by observation. We find, for example, that it is 1 o'clock by our chronometer, when it is 2 o'clock and 30 minutes at the place of observation. Hence, the longitude is $15 \times 1.5 = 22\frac{1}{2}°$ E. Had we traveled westward until our chronometer was an hour and a half in advance of the time at the place of observation (that is, so much later in the day), our longitude would have been $22\frac{1}{2}°$ W. But it would not be necessary to repair to London in order to set our chronometer to Greenwich time. This might be done at any observatory, or any place whose longitude had been accurately determined. For example, the time at New York is 4h. 56m. 4s.5 behind that of Greenwich. If, therefore, we set our chronometer so much before the true time at New York, it will indicate the time at Greenwich. Moreover, on arriving at different places, anywhere on the earth, whose longitude is accurately known, we may learn whether our chronometer keeps accurate time or not; and if not, the amount of its error. Chronometers have been constructed of such an astonishing degree of accuracy, as to deviate but a few seconds in a voyage from London to Baffin's Bay and back, during an absence of several years. But chronometers which are sufficiently accurate to be depended on for long voyages, are too expensive for general use, and the means of verifying their accuracy are not sufficiently easy. Moreover, chronometers, by being transported from one place to another, change their daily rate, or depart from that mean rate which they preserve at a fixed station. A chronometer, therefore, can not be relied on for determining the longitudes of places where the greatest degree of accuracy is required, especially where the instrument is conveyed over land, although the uncertainty attendant on one instrument may be nearly obviated by employing several, and taking their mean results.*

275. *Eclipses of the sun and moon* are sometimes used for determining the longitude. The exact instant of immersion or of emersion, or any other definite moment of the eclipse which presents itself to two distant observers, affords the means of comparing their difference of time, and hence of determining their difference of longitude. Since the entrance of the moon into the earth's shadow, in a lunar eclipse, is seen at the same instant of absolute time at all places where the eclipse is visible (Art. 262), this observation would be a very suitable one for finding the longitude, were it not that, on account of the increasing darkness of the penumbra near the boundaries of the shadow, it is difficult to determine the precise instant when the moon enters the shadow. By taking observations on the immersions of known spots on the lunar disk, a mean result may be obtained which will give the longitude with tolerable accuracy. In an eclipse of the sun, the instants of immersion and emersion may be observed with greater accuracy, although, since these do not take place at the same instant of absolute time, the calculation of the longitude from observations on a solar eclipse is complicated and laborious.

A method very similar to the foregoing, by observations on eclipses of Jupiter's satellites, and on occultations of stars, will be mentioned hereafter.

---

* Woodhouse, p. 838.

276. *The Lunar method of finding the longitude*, at sea, is in many respects preferable to every other. It consists in measuring (with a sextant) the angular distance between the moon and the sun, or between the moon and a star, and then turning to the Nautical Almanac,* and finding what time it was at Greenwich when that distance was the same. The moon moves so rapidly, that this distance will not be the same except at very nearly the same instant of absolute time. For example, at 9 o'clock, A. M., at a certain place, we find the angular distance of the moon and the sun to be 72°; and on looking into the Nautical Almanac, we find that the time when this distance was the same for the meridian of Greenwich was 1 o'clock, P. M.; hence we infer that the longitude of the place is four hours, or 60° west.

The Nautical Almanac contains the true angular distance of the moon from the sun, from the four large planets (Venus, Mars, Jupiter, and Saturn), and from nine bright fixed stars, for the beginning of every third hour of mean time for the meridian of Greenwich; and the time corresponding to any intermediate distance, may be found by proportional parts.†

277. It would be a very simple operation to determine the longitude by Lunar Distances, if the process, as described in the preceding article, were all that is necessary; but the various circumstances of parallax, refraction, and dip of the horizon, would differ more or less at the two places, even were the bodies (whose distances were taken) in view from both, which is not necessarily the case. The observations, therefore, require to be reduced to the center of the earth, being cleared of the effects of parallax and refraction. Hence, three observers are necessary in order to take a lunar distance in the most exact manner, viz., two to measure the altitudes of the two bodies respectively, at the same time that the third takes the angular distance between them. The altitudes of the two luminaries

* The *Nautical Almanac* is published annually three or four years in advance, containing all necessary tables and information, for the use of navigators. The English Board of Longitude have for a long series of years issued such a work. The *American Ephemeris and Nautical Almanac*, which possesses the same general character, was commenced but a few years since.

† See Bowditch's Navigator, tenth ed., p. 226.

at the time of observation must be known, in order to estimate the effects of parallax and refraction.

278. Since the invention of the magnetic telegraph, it has been employed to determine the differences of longitude between fixed stations on land with a precision which was before altogether unattainable. Suppose two stations to be connected by the telegraphic line, and that there is at each a clock keeping the local time. The observers agree beforehand at what time by his own clock the one at the most easterly station shall commence giving signals; and also at what time the other shall commence giving another series according to *his* clock. The interval between successive signals is also previously determined. When the moment arrives, the first observer strikes the telegraphic key at the exact beat of the clock, and the second observer records the *time* of the signal as shown by his own clock, and thus they continue to do till the full series is recorded. The second observer then commences sending signals, which are in like manner recorded by the first. The velocity of the electric current is so great that the *absolute* time of making a signal at one station, and of perceiving it at the other, may be considered identical, so that the difference which is indicated by the two clocks in each case is wholly due to difference of longitude. Still greater precision is attained by causing the signal key at each station to record its own movement on the line of second-marks made by the clock at the other station (Art. 125).

TIDES.

279. The tides are an alternate rising and falling of the waters of the ocean at regular intervals. They have a maximum and minimum twice a day; and the daily maximum and minimum reach their highest and lowest values twice in a synodical revolution of the moon, or 29½ days. The maximum of the daily tide is called *high tide*, and the minimum *low tide*. The maximum tide during a lunation is called the *spring tide*, the minimum, *neap tide*. The rising of the tide is called *flood*, and the falling, *ebb*.

Similar tides, whether high or low, occur on opposite sides

of the earth at once. Thus at the same time it is high tide at any given place, it is also high tide on the inferior meridian, and the same is true of the low tides.

The interval between two successive high tides is 12h. 25m.; or, if the same tide be considered as returning to the meridian, after having gone around the globe, its return is about 50 minutes later than it occurred on the preceding day. In this respect, as well as in various others, it corresponds very nearly to the motions of the moon.

The average height for the whole globe is about 2½ feet; or, if the earth were covered uniformly with a stratum of water, the difference between the two diameters of the oval would be 5 feet, or more exactly 5 feet and 8 inches; but its natural height at various places is very different, sometimes rising to 60 or 70 feet, and sometimes being scarcely perceptible. At the same place, also, the phenomena of the tides are very different at different times.

Inland lakes and seas, even those of the largest class, as Lake Superior, or the Caspian, have no perceptible tide.

280. *Tides are caused by the unequal attraction of the sun and moon upon different parts of the earth.*

Suppose the projectile force by which the earth is carried forward in her orbit to be suspended, and the earth to fall toward one of these bodies, the moon for example, in consequence of their mutual attraction. Then, if all parts of the earth fell equally toward the moon, no derangement of its different parts would result, any more than of the particles of a drop of water in its descent to the ground. But if one part fell faster than another, the different portions would evidently be separated from each other. Now this is precisely what takes place with respect to the earth in its fall toward the moon. The portions of the earth in the hemisphere next to the moon, on account of being nearer to the center of attraction, fall faster than those in the opposite hemisphere, and consequently leave them behind. The solid earth, on account of its cohesion, can not obey this impulse, since all its different portions constitute one mass, which is acted on in the same manner as though it were all collected in the center; but the waters on the surface, moving freely under this impulse, endeavor to desert the solid mass and fall toward the moon. For a similar reason the waters in the opposite hemisphere falling less toward the moon than the solid earth, are left behind, or appear to rise from the center of the earth.

281. Let DEFG (Fig. 56) represent the globe; and, for the sake of illustrating the principle, we will suppose the waters entirely to cover the globe at a uniform depth. Let *defg* represent the solid globe, and the circular ring exterior to it, the covering of waters. Let C be the center of gravity of the solid mass, A that of the hemisphere next to the moon, and B that of the remoter hemisphere. Now the force of attraction exerted by the moon acts in the same manner as though the solid mass were all concentrated in C, and the waters of each hemisphere at A and B respectively; and (the moon being supposed above E) it is evident that A will tend to leave C, and C to leave B behind. The same must evidently be true of the respective portions of matter, of which these points are the centers of gravity. The waters of the globe will thus be reduced to an oval shape, being elongated in the direction of that meridian which is under the moon, and flattened in the intermediate parts, and most of all at points ninety degrees distant from that meridian.

Fig. 56.

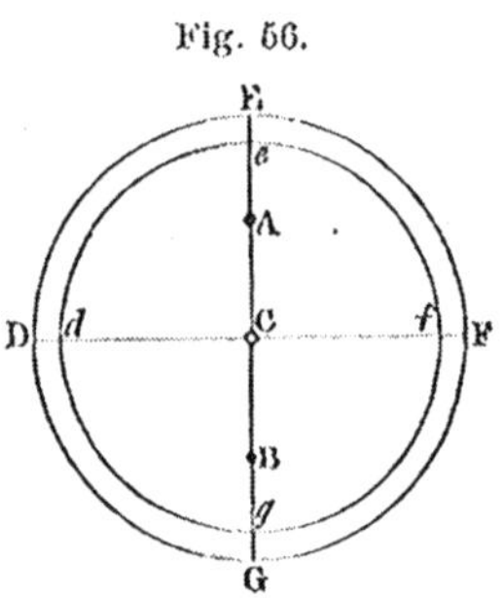

Were it not, therefore, for impediments which prevent the force from producing its full effects, we might expect to see the great *tide-wave*, as the elevated crest is called, always directly beneath the moon, attending it regularly around the globe. But the inertia of the waters prevents their instantly obeying the moon's attraction, and the friction of the waters on the bottom of the ocean, still further retards its progress. It is not therefore until several hours (differing at different places) after the moon has passed the meridian of a place, that it is high tide at that place.

282. The *sun* has a similar action to the moon, but only *one-third* as great. On account of the great mass of the sun compared with that of the moon, we might suppose that his action in raising the tides would be greater than the moon's; but the nearness of the moon to the earth more than compensates for the sun's greater quantity of matter. Let us, however, form a just conception of the advantage which the moon derives from her proximity. It is not that her actual amount of attraction is thus rendered greater than that of the sun; but it is that her attraction for the *different parts* of the earth is very unequal, while that of the sun is nearly uniform. It is the *inequality* of this action, and not the absolute force, that produces the tides. The diameter of the earth is $\frac{1}{30}$ of the distance of the moon, while it is less than $\frac{1}{10000}$ of the distance of the sun.

283. Having now learned the general cause of the tides, we will next attend to the explanation of *particular phenomena.*

The *Spring tides*, or those which rise to an unusual height twice a month, are produced by the sun and moon's acting in a line; and the *Neap tides*, or those which are unusually low twice a month, are produced by the sun and moon's acting 90 degrees from each other. The Spring tides occur at the syzygies; the Neap tides at the quadratures. At the time of new moon, the sun and moon both being on the same side of the earth, and acting upon it in the same line, their actions conspire, and the sun may be considered as adding so much to the force of the moon. We have already explained how the moon contributes to raise a tide on the opposite side of the earth. But the sun as well as the moon raises its own tide-wave, which, at new moon, coincides with the lunar tide-wave. At full moon also, the two luminaries conspire in the same way to raise the tide; for we must recollect that each body contributes to raise the tide on the opposite side of the earth as well as on the side nearest to it. At both the conjunctions and oppositions, therefore, that is, at the syzygies, we have unusually high tides. But here also the maximum effect is not at the moment of the syzygies, but 36 hours afterward.

At the quadratures, the solar wave is lowest where the lunar wave is highest; hence the low tide produced by the sun is subtracted from high water and produces the Neap tides. Moreover, at the quadratures the solar wave is highest where the lunar wave is lowest, and hence is to be added to the height of low water at the time of Neap tides. Hence the difference between high and low water is only about half as great at Neap tide as at Spring tide.

284. The power of the moon or of the sun to raise the tide is found by the doctrine of universal gravitation to be *inversely as the cube of the distance.** The variations of distance in the sun are not great enough to influence the tides very materially, but the variations in the moon's distances have a striking effect. The tides which happen when the moon is in perigee, are considerably greater than when she is in apogee; and if she happens to be in perigee at the time of the syzygies, the Spring tide is unusually high.

285. The declinations of the sun and moon cause the im mediate tide, at a given latitude, to be greater than the opposite one, or the reverse, according as the declination and latitude are alike or unlike. When the declination is nothing, both tides are alike at every place. For if the moon is in the plane

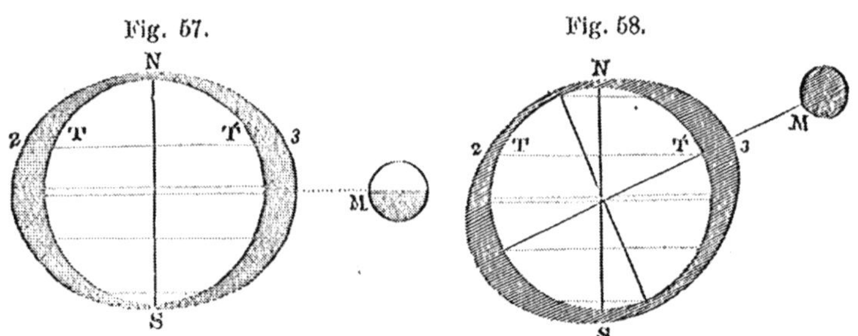

of the equator (Fig. 57),† then the highest points of both tide-waves are also on the equator; and, at a given latitude, a place by the earth's rotation is carried round through T2, T3, at

* La Place, Syst. du Monde, l. iv., c. x.

† Diagrams like these are apt to mislead the learner, by exhibiting the protuberance occasioned by the tides as much greater than the reality. We must recollect that it amounts, at the highest, to only a very few feet in eight thousand miles. Were the diagram, therefore, drawn in just proportions, the alterations of figure produced by the tides would be wholly insensible.

equal distances from the highest points, and therefore the immediate tide T3, and the opposite one T2, are equal; both are less, the higher the latitude, and at the poles there is no tide. When the moon is on the north side of the equator, as in figure 58, at her greatest northern declination, a place describing the parallel TT′ will have T3 for the height of the tide when the moon is on the superior meridian, and T2 for the height when the moon is on the inferior meridian. Therefore, all places north of the equator will have the highest tide when the moon is above the horizon, and the lowest when she is below it; the difference of the tides diminishing toward the equator, where they are equal. At the same time, places south of the equator have the highest tides when the moon is below the horizon, and the lowest when she is above it. When the moon is at her greatest declination, the highest tides will take place toward the tropics. The circumstances are all reversed when the moon is south of the equator.*

286. The motion of the tide wave, it should be remarked, is not a *progressive* motion, but a mere undulation, and is to be carefully distinguished from the currents to which it gives rise. If the ocean enveloped the earth, and the sun and moon were at rest in the equator, the tide-wave would travel at the same rate as the earth on its axis. Indeed, the correct way of conceiving of the tides, is first to regard the moon as at rest, and the earth as revolving and bringing successive parts under it, which parts are thus elevated in succession; and then, to consider the moon and tides as moving east 13° per day; thus making the time of the relative revolution of the tides westward, near 25 instead of 24 hours.

287. The tides of *rivers, narrow bays, and shores far from the main body of the ocean*, are not produced in those places by the direct action of the sun and moon, but are subordinate waves propagated from the great tide-wave.

Lines drawn through all the adjacent parts of any tract of water, which have high water at the same time, are called *cotidal lines.*† We may, for instance, draw a line through all

* Edin. Encyc., Art. *Astronomy*, p. 623.
† Whewell, Phil. Transactions for 1833, p. 148.

places in the Atlantic Ocean which have high tide on a given day at 1 o'clock, and another through all places which have high tide at 2 o'clock. The cotidal line for any hour may be considered as representing the summit or ridge of the tide-wave at that time; and could the spectator, detached from the earth, perceive the summit of the wave, he would see it traveling round the earth in the open ocean once in twenty-five hours, followed by another in twelve and a half hours, both sending branches into rivers, bays, and other openings into the main land. These latter are called *Derivative* tides, while those raised directly by the action of the sun and moon, are called *Primitive* tides.

288. The velocity with which the wave moves will depend on various circumstances, but principally on the depth, and probably on the regularity of the channel. If the depth be nearly uniform, the cotidal lines will be nearly straight and parallel. But if some parts of the channel are deep, while others are shallow, the tide will be detained by the greater friction of the shallow places, and the cotidal lines will be irregular. The direction, also, of the derivative tide may be totally different from that of the primitive. Thus (Fig. 59), if the great tide-wave, moving from east to west, be represented by the lines 1, 2, 3, 4; the derivative tide, which is propagated up a river or bay, will be represented by the cotidal lines 3, 4, 5, 6, 7. Advancing faster in the channel than next the banks, the tides will lag behind toward the shores, and the cotidal lines will take the form of curves, as represented in the diagram.

Fig. 59.

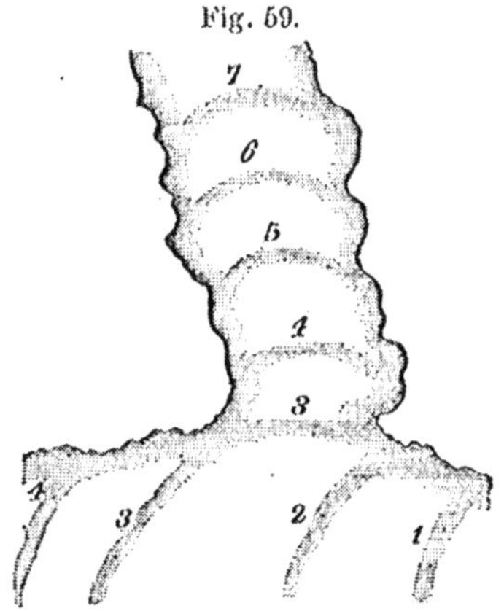

289. On account of the retarding influence of shoals, and an uneven, indented coast, the tide-wave travels more slowly along the shores of an island than in the neighboring sea, assuming convex figures at a little distance from the island and on opposite sides of it. These convex lines sometimes meet and become blended in such a manner as to create singular anomalies in a sea much broken by islands, as well as on coasts indented with numerous bays and rivers.* Peculiar phenomena are also produced, when the tide flows in at opposite extremities of a reef or island, as into the two opposite ends of Long Island Sound. In certain cases a tide-wave is forced into a narrow arm of the sea, and produces very remarkable tides. The tides of the Bay of Fundy (the highest in the world) sometimes rise to the height of 60 or 70 feet; and the tides of the river Severn, near Bristol, in England, rise to the height of 40 feet.

290. The *Unit of Altitude* of any place is the height of the maximum tide after the syzygies (Art. 283), being usually about 36 hours after the new or full moon. But as the amount of this tide would be affected by the distance of the sun and moon from the earth (Art. 284), and by their declinations (Art. 285), these distances are taken at their mean value, and the luminaries are supposed to be in the equator; the observations being so reduced as to conform to these circumstances. The unit of altitude can be ascertained by observation only. The actual rise of the tide depends much on the strength and direction of the wind. When high winds conspire with a high flood tide, as is frequently the case near the equinoxes, the tide rises to a very unusual height. We subjoin, from the American Almanac, a few examples of the unit of altitude for different places.

	Feet.
Cumberland, head of the Bay of Fundy, .	71
Boston, ...	11¼
New Haven, ...	8
New York, ...	5
Charleston, S. C., ...	6

291. The *Establishment* of any port is the mean interval between noon and the time of high water, on the day of new

* See an excellent representation and description of these different phenomena by Professor Whewell, Phil. Trans., 1833, p. 153.

or full moon. As the interval for any given place is always nearly the same, it becomes a criterion of the retardation of the tides at that place. On account of the importance to navigation of a correct knowledge of the tides, the British Board of Admiralty, at the suggestion of the Royal Society, recently issued orders to their agents in various important naval stations, to have accurate observations made on the tides, with the view of ascertaining the establishment and various other particulars respecting each station;* and the government of the United States is prosecuting similar investigations respecting our own ports.

292. According to Professor Whewell,† the tides on the coast of North America are derived from the great tide-wave of the South Atlantic, which runs steadily northward along the coast to the mouth of the Bay of Fundy, where it meets the northern tide-wave flowing in the opposite direction. Hence he accounts for the high tides of the Bay of Fundy.

293. The largest *lakes* and *inland seas* have no perceptible tides. This is asserted by all writers respecting the Caspian and Euxine, and the same is found to be true of the largest of the North American Lakes, Lake Superior.‡

Although these several tracts of water appear large, when taken by themselves, yet they occupy but small portions of the surface of the globe, as will appear evident from the delineation of them on an artificial globe. Now we must recollect that the primitive tides are produced by the *unequal* action of the sun and moon upon the different parts of the earth; and that it is only at points whose distance from each other bears a considerable ratio to the whole distance of the sun or the moon, that the inequality of action becomes manifest. The space required is larger than either of these tracts of water. It is obvious, also, that they have no opportunity to be subject to a derivative tide.

294. To apply the theory of universal gravitation to all

* Lubbock, Report on the Tides, 1833. † Phil. Trans., 1833, p. 172.
‡ See Experiments of Gov. Cass, Am. Jour. Science.

the varying circumstances that influence the tides, becomes a matter of such intricacy, that La Place pronounces "the problem of the tides" the most difficult problem of celestial mechanics.

295. The *Atmosphere* that envelops the earth must evidently be subject to the action of the same forces as the covering of waters, and hence we might expect a rise and fall of the barometer, indicating an atmospheric tide corresponding to the tide of the ocean. La Place has calculated the amount of this aerial tide. It is too inconsiderable to be detected by changes in the barometer, unless by the most refined observations. Hence it is concluded that the fluctuations produced by this cause are too slight to affect meteorological phenomena in any appreciable degree.*

CHAPTER IX.

OF THE PLANETS—INFERIOR PLANETS, MERCURY AND VENUS.

296. The name planet signifies a *wanderer*,† and is applied to this class of bodies because they shift their positions in the heavens, whereas the fixed stars apparently always maintain the same places with respect to each other. The planets known from a high antiquity, are Mercury, Venus, Earth, Mars, Jupiter, and Saturn. To these, in 1781, was added Uranus‡ (or *Herschel*, as it was formerly called, from the name of its discoverer), and, as late as 1846, another large planet, Neptune, was added to the list, making eight in all of those bodies usually called planets. Besides these, there is, between Mars and Jupiter, a remarkable group of small planets, called *Asteroids*, or more properly, *Planetoids*. Four of them were discovered near the beginning of the present century. From

* Bowditch's La Place, ii., p. 797.

† From the Greek, πλανητης.

‡ From Ουρανος.

that time till 1845, none were added; but since 1845, scarce a year has passed in which one or more have not been discovered; and sometimes five or six have been found in a single year. Though names have generally been given them, they are more commonly designated by a *number*, in the order of their discovery, the number being inclosed in a circle, which is intended to represent the planetary disk.

The foregoing are called *primary* planets. Several of these have one or more attendants, or satellites, which revolve around them as they revolve around the sun. The Earth has one satellite, namely, the moon; Jupiter has four; Saturn, eight; Uranus, six;* and Neptune, one. These bodies, also, are planets, but in distinction from the others, they are called *secondary* planets. Grouping the planetoids together, and giving them the rank of one primary planet (though all united would make but a small one), there are *nine* primaries and *twenty* secondaries.

297. The primary planets all (with the exception of the asteroids) have their orbits nearly in the same plane, and are never seen far from the ecliptic. Mercury, whose orbit is most inclined of all, never departs further from the ecliptic than about 7°, while most of the other planets pursue very nearly the same path with the earth, in their annual revolution around the sun. The asteroids, however, make wider excursions from the plane of the ecliptic, amounting, in the case of Pallas, to 34½°.

298. Mercury and Venus are called *inferior* planets, because their orbits are nearer to the sun than that of the earth; while all the others being more distant from the sun than the earth, are called *superior* planets. The planets present great diversities among themselves in respect to distance from the sun, magnitude, time of revolution, and density. They differ also in regard to satellites, of which, as we have seen, the Earth and Neptune have each one, Jupiter has four, Saturn eight, and Uranus six; while Mercury, Venus, and Mars, have

* Respecting the number of satellites belonging to Uranus, there is some doubt, which will be considered under the history of that planet.

none at all. It will aid the memory, and render our view of the planetary system more clear and comprehensive, if we classify, as far as possible, the various particulars comprehended under the foregoing heads.

299. DISTANCES FROM THE SUN.*

1. Mercury,	37,000,000	0.3870981
2. Venus,	68,000,000	0.7233316
3. Earth,	95,000,000	1.0000000
4. Mars,	145,000,000	1.5236923
5. Planetoids,	250,000,000	2.6612885
6. Jupiter,	495,000,000	5.2027760
7. Saturn,	900,000,000	9.5387861
8. Uranus,	1,800,000,000	19.1823900
9. Neptune,	2,800,000,000	30.0318000

The *dimensions* of the planetary system are seen from this table to be vast, comprehending a circular space nearly six thousand millions of miles in diameter. A railway car, traveling night and day at the rate of 20 miles an hour, and of course making 480 miles a day, would require about 50 days to travel round the Earth on a great circle, and about 500 days to reach the moon; but it will give some idea of the vastness of the planetary spaces to reflect, that setting out from the sun, and traveling from planet to planet at the same rate, to reach Mercury would require about 200 years; Venus, nearly 400; the Earth, 542; Mars, more than 800; Jupiter, toward 3,000; Saturn, above 5,000; Uranus, 10,000; Neptune, more than 16,000; and to cross the entire orbit of Neptune would require upward of 32,000 years.

It may aid the memory to remark, that in regard to the planets nearest the sun, the distances increase in an arithmetical ratio, while those most remote increase in a geometrical ratio. Thus, if we add 30 to the distance of Mercury, it gives us nearly that of Venus; 30 more gives that of the Earth; while Saturn is nearly twice the distance of Jupiter, and

* The distances in miles, as expressed in the first column, are to be treasured up in the memory, while the second column expresses the *relative* distance, that of the Earth being 1, from which a more exact determination may be made when required, the Earth's distance being taken at 95,298,260 miles.

Uranus twice that of Saturn. If this, however, were a perfectly correct rule, Neptune would be twice as far from the sun as Uranus, and therefore 3,600 millions of miles, whereas its actual distance is short of 3,000 millions. Between the orbits of Mars and Jupiter a great chasm appeared, which broke the continuity; but the discovery of the planetoids has filled the void. A more exact law of the series is that called *Bode's* law. It is as follows: if we represent the distance of Mercury by 4, and increase the following terms by the product of 3 into the ascending powers of 2, we shall obtain the relative distances of the planets from the sun. Thus,

Mercury, ...	4	= 4
Venus, ...	$4+3.2^0$	= 7
Earth, ...	$4+3.2^1$	= 10
Mars, ...	$4+3.2^2$	= 16
Planetoids, ...	$4+3.2^3$	= 28
Jupiter, ...	$4+3.2^4$	= 52
Saturn, ...	$4+3.2^5$	=100
Uranus, ...	$4+3.2^6$	=196
Neptune, ...	$4+3.2^7$	=388

Before the discovery of Neptune, Bode's law rudely expressed the relative distances from the sun; but it signally fails of including the new planet, as it gives near 3,600 millions instead of the true distance, 2,800 millions.

But the relative distances from the sun are accurately obtained by Kepler's third law,—that the squares of the periodic times are as the cubes of the distances (Art. 171). Thus the Earth's distance being previously ascertained by means of the sun's horizontal parallax (Art. 87), and the period of any other planet, as Jupiter, being learned from observation, we may say, as the square of the Earth's period (365.256 days) is to the square of Jupiter's period (4332.586 days), so is the cube of the Earth's distance to the cube of Jupiter's distance, the cube root of which will be the distance itself. Or, to express the same truth more concisely, $365.256^2 : 4332.586^2 :: 1^3 : 5.202^3$. Of course, this method can as yet be used only approximately for Neptune, which has described but a short portion of its orbit since its discovery.

300. MAGNITUDES.

	Diameter in Miles.	Mean apparent Diameter.	Mass.	Volume.
Mercury, ...	2,950*	8″.	8	$\frac{1}{19}$
Venus, ...	7,800	17″.	97	$\frac{9}{10}$
Earth, ...	7,912		100	1
Mars, ...	4,500	6″.	15	$\frac{1}{6}$
Planetoids, ...		0″.5	uncertain.	
Jupiter, ...	89,000	37″.	37,171	1,400
Saturn, ...	79,000	16″.	11,121	1,000*
Uranus, ...	35,000	4″.	1,564	86
Neptune, ...	31,000*	2″.5	672	60

Diagrams and orreries, as usually constructed, wholly fail of giving any just conceptions of the distances of the planets from the sun and from each other. If we represent, for instance, the distance of the earth by 1 foot, we shall require 30 feet in order to reach the place of Neptune; and when we have constructed a diagram on so large a scale, we must still bear in mind that each foot represents a space of nearly 100 millions of miles.†

We remark here a great diversity in regard to magnitude—a diversity which does not appear to be subject to any definite law. While Venus, an inferior planet, is nine-tenths as large as the earth, Mars, a superior planet, is only one-sixth, while Jupiter is fourteen hundred times as large. Although several of the planets, when nearest to us, appear brilliant and large when compared with the fixed stars, yet the angle which they subtend is very small, that of Venus, the greatest of all, never exceeding about 1′, or more exactly 61″.2, and that of Jupiter, when greatest, being only about $\frac{3}{4}$ of a minute.

* Hind.

† For the purposes of illustration to a class or to a popular audience, the following plan of representation is recommended, not only for the entire solar system, but for each of the subordinate systems, as that of Jupiter or Saturn. Stretch upon the wall a piece of black cambric, as long as the room will allow, say 30 ft. This length may be taken for the radius of Neptune's orbit. At one end, attach a circle of white cloth, $\frac{1}{4}$ inch in diameter for the sun. From it as a center describe arcs across the cloth at proper distances for the several orbits, and sew white tape on these arcs. We then have the planetary distances, and the size of the sun, upon one scale. The planets themselves can not be represented, since the largest of them would be almost invisibly small. The cloth may be conveniently rolled up, when not in use.

The distance of one of the near planets, as Venus or Mars, may be determined from its parallax; and the distance being known, its real diameter can be estimated from its apparent diameter, in the same manner as we estimate the diameter of the sun. (Art. 145.)

301. PERIODIC TIMES.

	Sidereal revolution.			Mean daily motion.
Mercury,	3 months,	or	88 days,	4° 5′ 32″.6
Venus,	7½ "	"	224 "	1° 36′ 7″.8
Earth,	1 year,	"	365 "	0° 59′ 8″.3
Mars,	2 "	"	687 "	0° 31′ 26″.7
Ceres,	4½ "	"	1,681 "	0° 12′ 50″.9
Jupiter,	12 "	"	4,332 "	0° 4′ 59″.3
Saturn,	29 "	"	10,759 "	0° 2′ 0″.6
Uranus,	84 "	"	30,686 "	0° 0′ 42″.4
Neptune,	164½ "	"	60,127 "	0° 0′ 21″.5

From this view it appears that the planets nearest the sun move most rapidly. Thus Mercury performs nearly 350 revolutions while Uranus performs one. This is evidently not owing merely to the greater dimensions of the orbit of Uranus, for the length of its orbit is not 50 times that of the orbit of Mercury, while the time employed in describing it is 350 times that of Mercury. Indeed, this ought to follow from Kepler's law, that the *squares* of the periodic times are as the *cubes* of the distances; from which it is manifest that the times of revolution increase faster than the dimensions of the orbit. Accordingly, the apparent progress of the most distant planets is exceedingly slow, the rate of Uranus being only 42″.4 per day; so that for weeks and months, and even years, this planet but slightly changes its place among the stars.

The planets are divided into two classes: first, the *inferior*, which have their orbits nearer to the sun than that of the earth; and secondly, the *superior*, which have their orbits exterior to the earth's orbit.

THE INFERIOR PLANETS, MERCURY AND VENUS.

302. The inferior planets, Mercury and Venus, having their orbits far within that of the earth, appear to us as attendants upon the sun. Mercury never appears further from the sun than 29° (28° 48′), and seldom so far; and Venus never more than about 47° (47° 12′). Both planets, therefore, appear either in the west soon after sunset, or in the east a little before sunrise. In high latitudes, where the twilight is prolonged, Mercury can seldom be seen with the naked eye, and then only at the periods of its greatest elongation.* The reason of this will readily appear from the following diagram.

Let S represent the sun, E the earth, and MN Mercury at its greatest elongations from the sun, and OZP a portion of the sky. Then, since we refer all distant bodies to the same concave sphere of the heavens, we should see the sun at Z and Mercury at O, when at its greatest eastern elongation, and at P when at its greatest western elongation; and while passing from M to N through Q, it would appear to describe the arc OP; and while passing from N to M, through R, it would appear to run back across the sun on the same arc. It is further evident that it would be visible only when at or near one of its greatest elongations; being at all other times so near the sun as to be lost in his light.

Fig. 60.

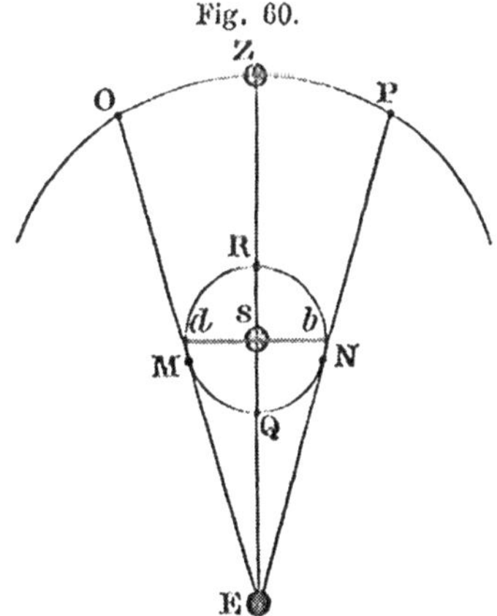

303. A planet is said to be in *conjunction* with the sun when it is seen in the same part of the heavens with the sun,

* Copernicus is said to have lamented, on his death-bed, that he had never been able to obtain a sight of Mercury; and Delambre, a great French astronomer, saw it but twice.

or when it has the same longitude. Mercury and Venus have each two conjunctions, the inferior and the superior. The *inferior conjunction* is its position when in conjunction on the same side of the sun with the earth, as at Q in the figure; the *superior conjunction* is its position when on the side of the sun most distant from the earth, as at R.

304. The period occupied by a planet between successive conjunctions of the same kind is called its *synodical revolution.* Both the planet and the earth being in motion, the time of the synodical revolution exceeds that of the sidereal revolution of Mercury or Venus; for when the planet comes round to the place where it before overtook the earth, it does not find the earth at that point, but far in advance of it. Thus, let Mercury come into its inferior conjunction at Q (Fig. 60). In about 88 days the planet will come round to the same point again; but meanwhile the earth has moved forward through nearly a fourth part of her revolution, and will continue to move onward while Mercury, with a swifter motion, is following on to overtake her, the case being analogous to the hour and minute hand of a clock. Having the *sidereal* period of a planet, which may always be accurately determined by observation, we may ascertain its *synodical* period, as follows:

By the table in article 301, the mean daily motion of Mercury is $4° \, 5' \, 32''.6 = 14732''.6$, and that of the earth is $59' \, 8''.3 = 3548''.3$. Therefore, $14732''.6 - 3548.3 = 11184''.3$, which is the average gain of Mercury over the earth in a day. But in order to overtake the earth, Mercury must complete one revolution and as much of another as the earth has performed, until the planet overtakes it; that is, the planet must *gain* an entire revolution. Now,

$11184''.3 : 1$ day :: $360° : 115.8$ days, the *synodical period of Mercury.* In like manner, the daily gain of Venus is $2219''.5$, and

$2219''.5 : 1$ day :: $360° : 583.9$ days, the *synodical period of Venus.*

305. *The motion of an inferior planet is direct in passing through its superior conjunction, and retrograde in passing through its inferior conjunction.*

Thus Venus, while going from B through D to A (Fig. 61), moves in the order of the signs, or from west to east, and would appear to traverse the celestial vault B′ S′ A′ from right to left; but in passing from A through C to B, her course would be retrograde, returning on the same arc from left to right. *If the earth were at rest*, therefore (and the sun, of course, at rest), the inferior planets would appear to oscillate backward and forward across the sun. But it must be recollected that the earth is moving in the same direction with the planet, as respects the signs, but with a slower motion. This modifies the apparent motions of the planet, accelerating it in the superior, and retarding it in the inferior conjunctions. Thus, in figure 61, Venus, while moving through BDA, would

Fig. 61.

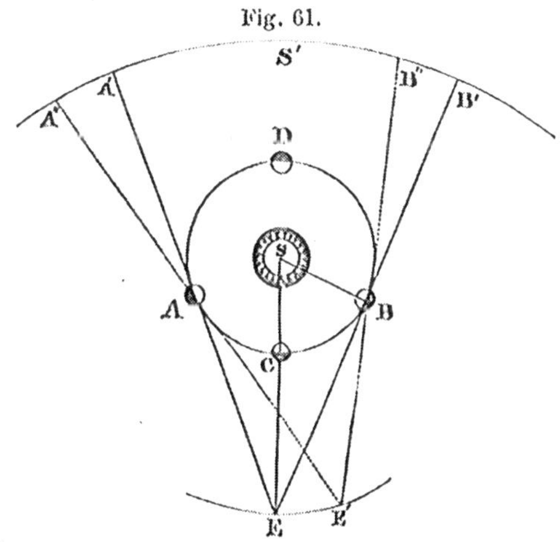

seem to move in the heavens from B′ to A′, were the earth at rest; but meanwhile the earth changes its position from E to E′, by which means the planet is not seen at A′, but at A″, being accelerated by the arc A′A″, in consequence of the earth's motion. On the other hand, when the planet is passing through its inferior conjunction ACB, it would appear to move backward in the heavens A′ to B′, if the earth were at rest, but from A′ to B″, if the earth has, in the mean time, moved from E to E′, being retarded by the arc B′B″. Although the motions of the earth have the effect to accelerate the planet in the superior, and to retard it in the inferior conjunction, yet on account of the greater distance, the apparent motion of the planet is much slower in the superior than in the inferior conjunction.

306. *In passing from either conjunction to the other, an inferior planet is stationary at a point a little way from the greatest elongation toward the inferior conjunction.*

If the earth, were at rest the stationary points would be at the greatest elongations A and B; for then the planet would be moving directly toward or from the earth, and would be seen for some time in the same place in the heavens; but the earth itself is moving nearly at right angles to the line of the planet's motion, and therefore a direct apparent motion is given to the planet. Hence we need to choose such a position for the planet that its retrograde movement shall be just sufficient to counteract this. Of course it must be on the arc ACB. But, as the planet's angular velocity is much greater than the earth's, it must be near A or B, where the motion is quite oblique to our own, else the retrogradation will be too rapid to neutralize the direct motion caused by the earth's progress. The stationary point for Mercury is at an elongation of 15° or 20° from the sun, that of Venus at about 29°.

307. *Mercury and Venus exhibit to the telescope phases similar to those of the moon.*

When on the side of their inferior conjunction, as from A to B through C (Fig. 61), these planets appear horned, like the moon in her first and last quarters; and when on the side of their superior conjunctions, as from B to A through D, they appear gibbous. At the moment of superior conjunction, the whole enlightened orb of the planet is turned toward the earth, and the appearance would be that of the full moon, but the planet is too near the sun to be commonly visible. All these changes of figure, resulting from the different positions of the planet with respect to the sun and earth, will be readily understood by inspecting the diagram (Fig. 61).

The phases show that these bodies are not self-luminous, but shine only as they reflect to us the light of the sun.; and all the planets in some way give evidence of the same fact.

308. *The distance of an inferior planet from the sun, may be found by observations at the time of its greatest elongation.*

Thus if E (Fig. 62) be the place of the earth, and C that of Venus at the time of her greatest elongation, the angle SCE will be known, being a right angle. The angle SEC is the greatest elongation; this is known by observation. Hence, if ES is considered to be known, Rad : sin E : : SE : SC, which is the distance of the planet from the sun. But if SE be not definitely known, then this proportion gives only the *ratio* of the distances of the earth and the inferior planet from the sun. In finding the earth's distance from the sun by means of the transit of Venus (Art. 318), this ratio will be employed. If the orbits were both circles, this method would be very exact; but being elliptical, we obtain the *mean* value of the radius SC by observing its greatest elongation in different parts of its orbit.*

Fig. 62.

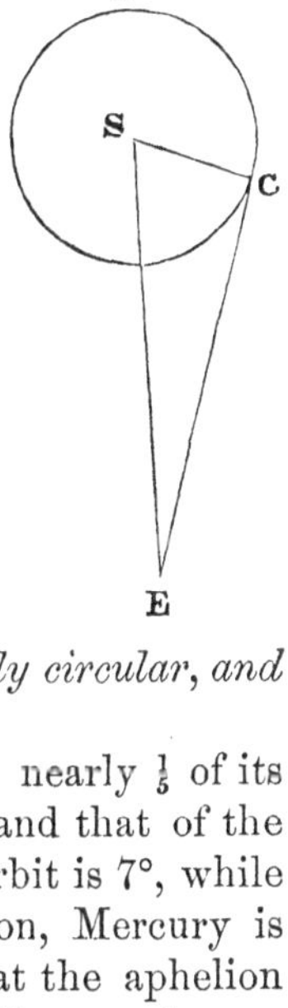

309. *The orbit of Mercury is more eccentric, and more inclined to the ecliptic, than that of any other planet;*† *while that of Venus is nearly circular, and but little inclined to the ecliptic.*

The *eccentricity* of the orbit of Mercury is nearly $\frac{1}{5}$ of its semi-major axis, while that of Venus is $\frac{1}{145}$; and that of the earth only $\frac{1}{59}$; the *inclination* of Mercury's orbit is 7°, while that of Venus is only $3\frac{1}{2}$°.‡ At the perihelion, Mercury is only 29 millions of miles from the sun, while at the aphelion his distance is 44 millions, a variation of 15 millions, and more than five times as great as in the case of the earth. On account of his different distances from the earth, Mercury is also subject to much variation in his apparent diameter, which is 12″ in perigee, but only 5″ in apogee.

310. After the mean distance has been found (Art. 307), the periodic time is obtained, by applying Kepler's third law to the orbit of the planet, and that of the earth. From this is calculated the synodical period (Art. 304).

* Herschel's Outlines, p. 275. † The asteroids are of course excepted.
‡ Baily's Tables.

The synodical period is also obtained very accurately from the interval between two transits across the sun's disk.

311. *An inferior planet is brightest at a certain point between its greatest elongation and inferior conjunction.*

As to *distance*, the planet would give us most light when nearest, *i. e.*, at the inferior conjunction; but so far as the *phase* is concerned, it would give us most at the superior conjunction, where the planet is at the full. Of course, the maximum is at some point between the two conjunctions; and by calculation it is found between the inferior conjunction and greatest elongation, within a few degrees of the latter. Venus appears most luminous when about 40° from the sun, and is sometimes visible all day.

312. *Mercury and Venus both revolve on their axes in nearly the same time with the earth.*

The diurnal period of Mercury is a little greater than that of the earth, being 24h. 5m. 28s., and that of Venus is a little less than the earth's, being 23h. 21m. 7s. The revolutions on their axes have been determined by means of some spot or mark seen by the telescope, as the revolution of the sun on his axis is ascertained by means of his spots.

313. Venus is regarded as the most beautiful of the planets, and is well known as the *morning and evening star*. The most ancient nations did not indeed recognize the evening and morning star as one and the same body, but supposed they were different planets, and accordingly gave them different names, calling the morning star Lucifer, and the evening star Hesperus. At her period of greatest splendor, Venus casts a shadow, and is sometimes visible in broad daylight. This occurred in a very striking manner in September, 1852, Venus being on the meridian about 9 o'clock, A. M., and her northern declination nearly 15 degrees. Although not 45° from the inferior conjunction, and consequently exposing only a portion of her disk, like that of the moon when three or four days old, yet her light is then estimated as equal to that of twenty stars of the first magnitude.* At her period of greatest elongation,

* Francœur, Uranography, p. 125.

Venus is visible from three to four hours after the setting, or before the rising of the sun.

314. *Every eight years, Venus forms her conjunctions with the sun in the same part of the heavens.*

The sidereal period of Venus being 224.7 days, and that of the earth 365.256 days, *thirteen* revolutions of Venus are accomplished in nearly the same time as *eight* revolutions of the earth: for 224.7×13=2921, and 365.256×8=2922. At the end, therefore, of 2922 days, or eight years, the two bodies will come round to the same point of the heavens, and be in the same situation in their respective orbits, as at the beginning. Consequently, whatever appearances of this planet arise from its positions with respect to the earth and the sun (as, for example, being visible in the daytime), they are repeated every eight years in nearly the same form.

TRANSITS OF THE INFERIOR PLANETS.

315. *The transit of Mercury or Venus, is its passage across the sun's disk, as the moon passes over it in a solar eclipse.*

As a transit takes place only when the planet is in inferior conjunction, at which time her motion is retrograde (Art. 305), it is always from left to right, and the planet is seen projected on the solar disk in a black round spot. Were the orbits of these planets coincident with the earth's orbit, a transit would occur at some part of the earth at every inferior conjunction, as there would be an eclipse of the sun at every new moon, were the moon's revolution in the plane of the ecliptic. But the orbit of Venus makes an angle of 3½° with that of the earth, and the orbit of Mercury an angle of 7°; and, moreover, the apparent diameter of each of these bodies is very small, both of which circumstances conspire to render a transit a comparatively rare occurrence, since it can happen only when the sun, at the time of an inferior conjunction, happens to be at, or extremely near the planet's node. The nodes of Mercury lie in that part of the earth's orbit which it passes in the months of May and November. It is only in these months, therefore, that transits of Mercury can occur. For a similar reason, those of Venus occur only in June and December. Since the nodes of both planets have a small retrograde motion, the months in which transits occur, will change in the course of ages; but the months for transits will for a long time remain the same as at present, since the nodes of Mercury change their places only 13′, and those of Venus only 31′ in a century.*

The first prediction of this phenomenon was made by Kepler, and was that of a transit of Mercury, which occurred on the 7th of November, 1631. As early as 1629, Kepler announced to astronomers that his tables gave the latitude of Mercury, at the conjunction which was to take place on that day, less than the sun's semi-diameter; consequently, that the planet, in passing by the sun, would be nearer the sun's center than the length of the sun's radius, and of course appear on his disk. The event corresponded to the prediction. The transit of Mercury, which occurred on the 8th of November, 1848, was the 25th since the one predicted by Kepler, averaging nearly one in 8 years, although they take place at very unequal intervals.

316. *The shortest interval between two successive transits of Mercury is 3½ years, and of Venus, 8 years:* but sometimes they are separated by long intervals, especially those of Venus. The only transits of Venus in the 19th century, are in 1874 and 1882; and in the 20th, not a single one will occur; the intervals being 8, 121½, 8, 105½, 8, 121½, &c., years. The intervals between the transits of Mercury, from 1848, through the century, are 13, 7, 9½, 3½, 9½, and 3½ years. The shortest interval for Mercury at the *same* node is 7 years; hence, at opposite nodes, two transits may occur within 3½ years. More of the transits of Mercury happen in November than in May, because the orbit of this planet (which has a great eccentricity, Art. 308), is so situated that in November the planet is near its perihelion, and is then more likely to be projected on the sun, in passing its inferior conjunction, than in a part of its orbit more distant from the sun.

Let us see how the *intervals* between the transits of Mercury

---

* Hind.

or Venus are found. Since Venus, for example, completes one revolution around the sun in 224.7 days, and the earth in 365.256, and since the number of times each will revolve in a given period is inversely as the time of one revolution, therefore, in 224,700 revolutions of the earth, and 365,256 revolutions of Venus, the two bodies would meet exactly at the same node as before. But 224,700 : 365,256 : : 8 : 13 *nearly;* so that transits of Venus are sometimes repeated at intervals of 8 years, and if the ratio of 8 to 13 were *exactly* that of the two first terms of the proportion, we should have a transit of Venus every 8 years. The ratio of 227 to 369 is still nearer that of those terms; and hence a transit after 227 years is still more probable; but since there are two nodes, the chance is doubled, so that a transit is highly probable after an interval of 113½ years. The two transits of Venus in the 18th century occurred in June, 1761, and June, 1769, 8 years apart; the two transits of the 19th century are in December, 1874, and December, 1882, the intervals being 105½ and 8 years. The average long interval is 113½ years, but it may be lengthened to 121½, or shortened to 105½, according as the preceding transit took place before or after passing the node.

317. The great interest attached by astronomers to a transit of Venus, arises from its furnishing the most accurate means in our power of determining the *sun's horizontal parallax*—an element of great importance, since it leads to a knowledge of the distance of the earth from the sun, and consequently, by the application of Kepler's third law (Art. 183), of the distances of all the other planets. Hence, in 1769, great efforts were made throughout the civilized world, under the patronage of different governments, to observe this phenomenon under circumstances the most favorable for determining the parallax of the sun. The method of finding the parallax of a heavenly body, described in Art. 85, can not be relied on to a greater degree of accuracy than 4″. In the case of the moon, whose greatest parallax amounts to about 1°, this deviation from absolute accuracy is not material, but it amounts to nearly half the entire parallax of the sun; and since the distance is inversely as the horizontal parallax, such an error would make the distance of the sun either twice as great, or only two-thirds as great as the true distance, according as the parallax was 4″ below or 4″ above the truth.

318. If the sun and Venus were equally distant from us, they would be equally affected by parallax as viewed by spectators in different parts of the earth, and consequently their *relative* situation would not be altered by such a difference in the points of view; but since Venus, at the inferior conjunction, is only about one-third as far off as the sun, her parallax is proportionally greater, and therefore spectators, at distant points, will see Venus projected on different parts of the solar disk; and as the planet traverses the disk, she will appear to describe chords of different lengths, by means of which the duration of the transit may be estimated at different places. The difference in the duration of the transit, as viewed from opposite parts of the earth, does not amount to many minutes; but to make it as large as possible, places very distant from each other are selected for observation. Thus, in the transit of 1769, among the places selected, two of the most favorable were Wardhus, in Lapland, and Otaheite (now written *Tahiti*), one of the Society Islands, in the South Pacific Ocean, to which place the celebrated Captain Cook was dispatched by the British government for the express purpose of observing the transit.

Although the exact determination of the sun's horizontal parallax by this method is a very complicated and difficult problem, yet the *principle* on which the process depends admits of an easy illustration. Let E (Fig. 63) be the earth, V Venus, and S the center of the sun. Suppose A and B two

Fig. 63.

observers at the extremities of that diameter of the earth which is perpendicular to the orbit of Venus. At a certain moment the spectator A will see Venus on the sun's disk at a, and the spectator B will see it at b; and since AV and BV may be considered as equal to each other, as also Vb and Va; therefore the triangles VAB and Vab are similar; and AV : aV :: AB : ab. But AB is known; also the ratio of AV : aV (Art. 308); hence ab, the distance between the points at which the two observers see V projected on the sun's disk, is known in miles. We need now to obtain the *angular* value of ab. The observers carefully note the instant when Venus touches the disk, at the beginning of the transit, and also at the end. Thus the *time* of making the transit, as seen by each observer, is accurately obtained. But since the angular motion per hour, both of the planet and of the sun, is known, this time of crossing the disk can be changed into an arc; and we thus have the number of minutes of a degree in the chord cd, and also the number in ef, and of course in their halves ca and eb. But cS and eS, the angular semi-diameter of the sun at the same time is known; hence, in the right-angled triangles cSa, eSb, we readily find the minutes in Sa, Sb, the difference between which is the angular value of ab. We have, therefore, ascertained what angle is subtended by a line of given length, when placed at the sun and viewed from the earth; or, which is the same thing, placed at the earth and viewed from the sun; and therefore we know what angle at the sun is subtended by the earth's semi-diameter, which is the sun's horizontal parallax.

The observers can not, probably, be at points diametrically opposite, nor can they remain stationary during the transit, on account of diurnal motion; therefore allowance must be made for these circumstances. The line ab will be more accurately measured, according as the transit occurs nearer to the edge of the sun's disk, because of the greater inequality in the length of the chords. The solar parallax is so small that several stations should be occupied, so as to obtain a number of independent results. The parallax of the sun, as measured in 1769, is 8″.5776.*

Venus when on the side of her inferior conjunction, and Mars when near his opposition, each comes comparatively near to the earth, and at these times exhibits a large horizontal parallax. That of Venus, especially, may be obtained with great accuracy when she is near her greatest elongation; and

* Delambre, t. ii; Vince, Complete Syst., vol. i.; Woodhouse, p. 751; Herschel's Outlines, p. 255.

since it is easy, by article 308, to determine, at that time, the *ratio* of her distance from the sun to the earth's distance, it is a matter of great interest to astronomy to have the parallax of Venus, when thus situated, accurately found. For this purpose, the government of the United States, in 1849, sent an expedition, under Lieutenant Gilliss, to Chili, in order to take observations on Mars and Venus, especially the latter, during 1850, 1851, and 1852, in concert with the Observatory at Washington. These observations seem to indicate that the solar parallax is a little less than stated above, probably nearer 8″.5 than 8″.6.*

319. During the transits of Venus over the sun's disk in 1761 and 1769, a sort of penumbral light was observed around the planet by several astronomers, which was thought to indicate an *atmosphere*. This appearance was particularly observable while the planet was coming on and going off the solar disk. The total immersion and emersion were not instantaneous; but as two drops of water when about to separate form a ligament between them, so there was a dark shade stretched out between Venus and the sun, and when the ligament broke, the planet seemed to have got about an eighth part of her diameter from the limb of the sun.† The presence of an atmosphere is also indicated by appearances of twilight and indications of a horizontal refraction.‡

Although no satellite has hitherto been discovered attending either Mercury or Venus, yet suspicions have, at different times, been entertained of a satellite belonging to Venus. None has been seen in any of the transits of Venus; and although the distance of the satellite (if one exists) from the primary might have been too great to be projected with the primary on the sun, yet its absence on each of these occasions has strengthened the belief of astronomers that no such satellite exists.

* See Astron. Journ., Oct., 1858. † Edin. Encyc., Art. *Astronomy*. ‡ Hind.

CHAPTER X.

OF THE SUPERIOR PLANETS—MARS, THE PLANETOIDS, JUPITER, SATURN, URANUS, AND NEPTUNE.

320. The Superior planets are distinguished from the Inferior, by being seen at all distances from the sun from 0° to 180°. Having their orbits exterior to that of the earth, they of course never come between us and the sun, that is, they have never any inferior conjunction like Mercury and Venus, but they are seen in superior conjunction and in opposition. Nor do they, like the inferior planets, exhibit to the telescope different phases; but, with a single exception, they always present the side that is turned toward the earth fully enlightened. This is owing to their great distance from the earth: for were the spectator to stand upon the sun, he would, of course, always have the illuminated side of each of the planets turned toward him; but so distant are all the superior planets except Mars, that they are viewed by us very nearly as they would be if we actually stood on the sun.

321. Mars is a small planet, his diameter being only about half that of the earth, or 4500 miles.* He also, at times, comes nearer to us than any other planet except Venus. His *mean* distance is 145,200,000 miles; but in consequence of the eccentricity of his orbit, the distance varies greatly, the difference between the perihelion and aphelion distances being 27,000,000 miles. Mars is always near the ecliptic, never varying from it 2°. He is distinguished from all the other planets by his deep red color and fiery aspect; but his brightness and apparent magnitude vary much at different times, being sometimes nearer to us than at others by the whole diameter of the earth's orbit, that is, by about 190,000,000 miles. When Mars is on the same side of the sun with the earth, or at his opposition, he comes within 50,000,000 miles of the earth, and, rising

* Hind.

about the time the sun sets, surprises us by his magnitude and splendor; but when he passes to the other side of the sun to his superior conjunction, he dwindles to the appearance of a small star, being then 240,000,000 miles from us. Thus, let M

Fig. 64.

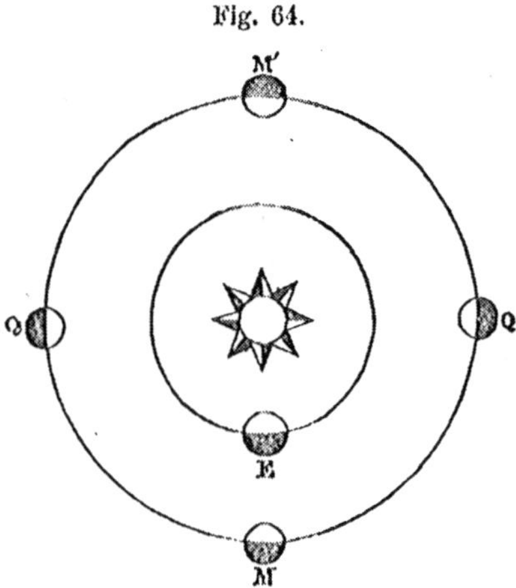

(Fig. 64) represent Mars in opposition, and M′ in superior conjunction, it is obvious that the planet must be nearer to us in the former situation than in the latter by the whole diameter of the earth's orbit.

322. Mars is the only one of the superior planets which exhibits phases. When he is toward the quadratures at Q or Q′, it is evident from the figure that only a part of the circle of illumination is turned toward the earth, such a portion of the remoter part of it being concealed from our view as to render the form more or less gibbous.

323. When viewed with a powerful telescope, the surface of Mars appears diversified with numerous varieties of light and shade. The region around the poles is marked by white spots, which vary their appearance with the changes of the seasons in the planet. Hence Dr. Herschel conjectured that they are owing to ice or snow which occasionally accumulates and melts, according to the position of each pole with respect to the sun.* It has been common to ascribe the ruddy light of this planet to an extensive and dense atmosphere, which was supposed to be distinctly indicated by the gradual diminution of light observed in a star as it approached very near to the planet in undergoing an occultation; but more recent observations afford no such evidence of an atmosphere.† By observations on the spots, we learn that Mars revolves on his axis in very nearly the same time with the earth (24h. 39m. 21ˢ.3); and that the angle between his equator and the plane of his orbit is also nearly the same as between the earth's equator and the ecliptic, the former being 28° 42′, the latter 23° 28′, so that the changes of seasons on Mars must resemble our own.

No *satellite* has ever been discovered belonging to Mars, although being situated at a greater distance from the sun than our globe, it might seem more especially to need such a luminary to cheer its dark nights. As the diurnal rotation of Mars is performed in nearly the same time as the earth, we should expect a similar flattening of the poles. Such is the fact, and the ellipticity of Mars exceeds that of the earth, being about one-fiftieth,‡ while the earth's ellipticity is one three-hundredth. This difference in the conjugate diameters may be readily observed when the planet is in opposition, the whole enlightened disk being then presented to us.

324. Mars being comparatively near to us when on the same side of the sun with the earth, and the ratio of his distance from the sun to that of the earth being easily obtained, astronomers have sought by means of his parallax, as by that of Venus, to find the sun's horizontal parallax. But the method by observations on Venus, as described in Art. 318, is more to be relied on.

325. The Asteroids or PLANETOIDS compose a group of very small planets, indefinite in number, whose orbits lie beyond that of Mars, at the distance of about 250,000,000 miles from the sun. The discovery of them commenced with the beginning of the present century. Kepler had long before noticed a large interval between Mars and Jupiter, which seemed to

* Phil. Trans., 1784. † Sir James South, Phil. Trans., 1833. ‡ Hind.

break the continuity of the series; and about the close of the last century, Bode, of Berlin, showed that a series of numbers following a certain law (Art. 299) would express quite accurately the distances from the sun, even to Uranus, which had just been discovered, if only the vacancy between Mars and Jupiter were supplied. So strongly were astronomers impressed with the idea that a planet would be found between Mars and Jupiter, that, in hope of discovering it, an association was formed on the continent of Europe of twenty-four observers, who divided the sky into as many zones, one of which was allotted to each member of the association. The discovery of the first of these bodies was, however, made accidentally by Piazza, an astronomer of Palermo, on the 1st of January, 1801. It was shortly afterward lost sight of, on account of its proximity to the sun, and was not seen again until the close of the year, when it was rediscovered in Germany. Piazza called it *Ceres*, in honor of the tutelary goddess of Sicily, and her emblem, the sickle ⚳, has been adopted as the appropriate symbol. The difficulty of finding Ceres induced Dr. Olbers, of Bremen, to examine, with particular care, all the small stars that lie near her path, as seen from the earth; and, while prosecuting these observations, in March, 1802, he discovered another similar body, very nearly at the same distance from the sun, and resembling the former in many other particulars. The discoverer gave to this second planet the name of *Pallas*, choosing for its symbol the lance ⚴, the characteristic of Minerva.

326. The most surprising circumstance connected with the discovery of Pallas, was the existence of two planets at nearly the same distance from the sun, and apparently having a common node; a circumstance that indicated an identity of origin. On account of this singularity, Dr. Olbers was led to conjecture that Ceres and Pallas are only fragments of a larger planet which had formerly circulated around the sun at this distance, and been shattered by some great convulsion.

In 1804, near one of the nodes of Ceres and Pallas, a third planet was discovered. This was named *Juno*, and the character ⚵ was adopted for its symbol, representing the starry scepter of the goddess. In 1807, a fourth planet, *Vesta*, was discovered, and for its symbol the character ⚶ was chosen—an altar surmounted with a censer holding the sacred fire. It is one of the largest of the asteroids, and has sometimes been seen by the naked eye.

327. From 1807 to 1845, a period of nearly forty years, no more of these small planets were discovered, and, up to this time, by the asteroids were meant the four little planets already enumerated—Ceres, Pallas, Juno, and Vesta. Meanwhile, very accurate maps of the stars, including all up to the tenth magnitude, had been published, especially in the region of the zodiac, and astronomers scrutinized these with such extreme closeness, that any wanderer appearing among them was likely to be immediately detected. Since 1845, new ones have been added to the list nearly every year. At the beginning of 1850, the whole number known was 10; of 1855, 23; of 1860, 57. Though feminine mythological names have been applied to nearly all of them, yet they are better designated by a small circle inclosing a number which expresses the order of their discovery: thus Ceres is (1); Thetis, (17); Pandora, (55), &c.

They vary somewhat in their mean distances, and of course in their periods. Some of them have orbits more eccentric, and others more inclined to the ecliptic, than any of the larger planets. They are too small to be measured with any certainty; the largest, Ceres, is not estimated to be more than 160 miles in diameter. It is probable that all of them united would form but an inconsiderable planet. Some of them are attended by a nebulosity, which indicates that they have an extensive atmosphere. The different planetoids have been first discovered by observers of many countries—England, France, Germany, Italy, and America.

328. Jupiter is distinguished from all the other planets by his great magnitude. His diameter is 89,000 miles, being more than 11 times, and his volume more than 1400 times that of the earth. His figure is strikingly spheroidal, the equatorial exceeding the polar diameter in the ratio of 107 to 100,* which is 21 times as great as the earth's ellipticity. This flattening

* Herschel.

of the poles is indeed quite perceptible by the telescope, and is obvious to the eye in a correct drawing of the planet. (See Frontispiece.) Such a figure might naturally be expected from the rapidity of his diurnal revolution, which is accomplished in about 10 hours (9h. 55m. 21^s.3).*

A place on the equator of Jupiter must revolve 450 miles per minute, or 27 times as fast as a place on the terrestrial equator. The distance of Jupiter from the sun is 495,000,000 miles (495,817,000).† His plane of rotation is but slightly inclined to the plane of his orbit (only about 3°), and consequently his climate experiences but a slight change of seasons.

329. The view of Jupiter through a good telescope is one of the most magnificent and interesting spectacles among the heavenly bodies. The disk expands into a large and bright orb like the full moon; the spheroidal figure which theory assigns to revolving worlds is here palpably exhibited to the eye; across the disk, arranged in parallel stripes, are discerned several dusky bands, called *belts;* and four bright satellites, always in attendance, but ever varying their positions, compose a splendid retinue. Indeed, astronomers gaze with peculiar interest on Jupiter and his moons, as affording a miniature representation of the whole solar system; repeating, on a smaller scale, the same revolutions, and exemplifying, in a manner more within the compass of our observation, the same laws as regulate the entire assemblage of sun and planets.

330. The *Belts of Jupiter* are variable in their number and dimensions. With smaller telescopes only one or two are seen across the equatorial regions; but with more powerful instruments the number is increased, covering a great part of the disk. Occasionally these belts retain nearly the same form and positions for many months together, while at other times they undergo great and sudden changes, and in one or two instances, they have been observed to break up and spread themselves over the whole face of the planet. The prevailing opinion among astronomers in reference to the nature of these belts is, that they are produced by disturbances in the planet's atmos-

* Airy. † Hind.

phere, which occasionally render its dark body visible; and, as the belts are found to traverse the disk in lines uniformly parallel to Jupiter's equator, they are inferred to be connected with the rotation of the planet upon its axis, the great rapidity of which would naturally produce peculiarities in its atmospheric phenomena.

331. The *Satellites of Jupiter* may be seen with a telescope of very moderate powers. Even a common spy-glass will enable us to discern them. Indeed, being nearly equal in brilliancy to the smallest stars visible to the naked eye, a slight increase of optical power brings them into view; and some few persons, endowed with extraordinary powers of vision, have supposed that they saw one of these little bodies without the aid of instruments; but on applying the telescope it has been found that three of the satellites have approached so near together as to appear like one.* In the largest telescopes, they severally appear as bright as Sirius does to the naked eye. With such an instrument, the view of Jupiter with his moons and belts is truly a magnificent spectacle—a world within itself. As the orbits of the satellites do not deviate far from the plane of the ecliptic, and but little from the equator of the planet (which nearly coincides with the ecliptic), they are usually seen almost in a straight line extending across the central part of the disk. (See Frontispiece.)

332. Jupiter's satellites are distinguished from one another by the denominations of *first*, *second*, *third*, and *fourth*, according to their relative distances from the primary, the first being that which is nearest to him.† Their apparent motion is oscillatory, like that of a pendulum, going alternately from their greatest elongation on one side to their greatest elongation on the other, sometimes in a straight line, and sometimes in an elliptical curve, according to the different points of view in

* Hind. Rev. Mr. Stoddard, a graduate of Yale College, missionary to the Nestorians, has repeatedly seen one of these bodies with the naked eye, from Mount Seir, near Oroomiah. Mr. Stoddard is known to the author as an excellent observer, and his testimony on this point may be fully relied on.

† Mythological names were long since proposed for the satellites of Jupiter, viz., Io, Europa, Ganymede, Calisto; but the mode of designating them by numbers generally prevails.

which we observe them from the earth. Their motion is alternately direct and retrograde; they are sometimes stationary; and, in short, they exhibit in miniature all the phenomena of the planetary system. Various particulars of the system are exhibited in the following table, the diameters being in miles, and the distances being taken from the center of the primary.*

Satellite.	Diameter.	Distances.	Sidereal Revolution.
1	2,440	278,500	1d. 18h. 28m.
2	2,190	443,300	3 13 15
3	3,580	707,000	7 3 43
4	3,060	1,243,500	16 16 32

From this table we see that Jupiter's satellites are all larger than the moon, the second exceeding it by only 30 miles in diameter, the third by 1420 miles. The third, the largest of the whole, has still only $\frac{1}{25}$th the diameter of the primary. The greater distances also of these moons compared with ours, reduces their apparent size and light as seen from Jupiter. Thus the largest of them would exhibit to a spectator on the equator of the planet, a diameter of only 36′, which is only a little greater than that of the moon, while the smallest would appear only one-fourth as large. It is noticeable, that the satellites of Jupiter make very quick revolutions, when compared with the earth's moon, although they are all at greater distances from the primary than the moon from the earth; the furthest revolving in about $\frac{3}{5}$ of the moon's period, and the nearest more than 16 times as quick. This is because they are so powerfully attracted by the planet. To prevent their being drawn in from their circular orbits, a great projectile velocity is necessary.

333. The orbits of the satellites are nearly or quite circular, and deviate but little from the plane of the planet's equator, and of course are but slightly inclined to the plane of his orbit. They are, therefore, in a similar situation with respect to Jupiter as the moon would be with respect to the earth, if her orbit nearly coincided with the ecliptic, in which

* Hind.

case she would eclipse the sun every new moon, and be herself eclipsed every full moon.

334. The *eclipses of Jupiter's satellites*, in their general conception, are perfectly analogous to those of the moon, but in their details they differ in several particulars. Owing to the much greater distance of Jupiter from the sun, and its greater magnitude, the cone of its shadow is more than sixty times that of the earth, stretching off into space more than 55,000,000 miles. On this account, as well as on account of the little inclination of their orbits to that of their primary, the three inner satellites of Jupiter pass through the shadow and are totally eclipsed at every revolution. The fourth satellite, owing to the greater inclination of its orbit, sometimes, though rarely, escapes eclipse, and sometimes merely grazes the limits of the shadow, or suffers a partial eclipse.* These eclipses, moreover, are not seen by us, as is the case with those of the moon, from the center of their motion, but from a remote station, and one whose situation, with respect to the line of the shadow, is variable. This of course makes no difference in the *times* of the eclipses, but a very great difference in their visibility, and in their apparent situations with respect to the planet at the moment of their entering or quitting the shadow.

335. The eclipses of Jupiter's satellites present some curious phenomena, which will be best understood from a diagram. Let A, B, C (Fig. 65) represent the earth in different parts of its orbit, revolving from A, through D, to C and B. If a line, joining S and A, be produced to meet the concave sphere of of the heavens *xy*, it marks the place of opposition, while the earth is at A. Hence, Jupiter, in the figure, is represented *east* of the opposition. When the earth arrives at D, Jupiter is *in* opposition; and when at C, he is *west* of opposition. Remembering now, that the satellites revolve in the same direction as the earth, it is obvious that when Jupiter is *east* of opposition, the *immersions* are seen, but the *emersions* generally take place behind the planet and are not seen; so that the *eclipse*, in this position of the bodies, always precedes the

* Herschel's Ast., p. 285.

occultation. But if the earth passes beyond D, to C, then Jupiter is *west* of opposition; for the place of opposition would be in SC produced. And now, each satellite passes behind the planet, and thus suffers occultation before it reaches the

Fig. 65.

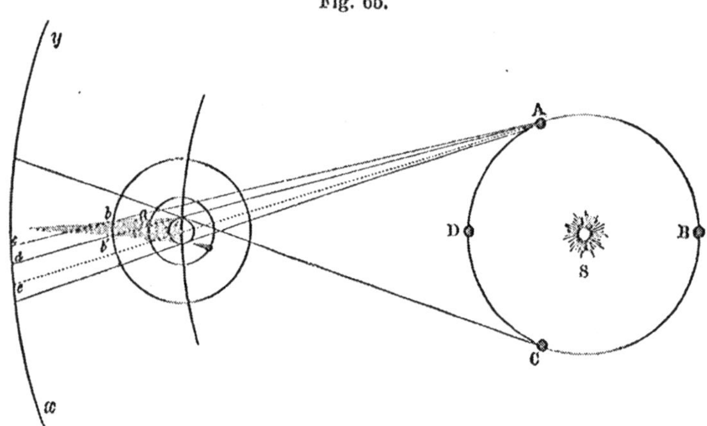

shadow, and is eclipsed; and, in *entering* the shadow, the satellites are generally behind the planet, while in *leaving* it, they are in sight; so that occultations precede eclipses. The fourth satellite often enters and leaves the shadow on the same side of the planet, and sometimes the third, and possibly the second, but this is never true of the first.

336. When one of the satellites is passing between Jupiter and the sun, it casts its shadow upon its primary, as the moon does on the earth in a solar eclipse, which is seen, by the telescope, traveling across the disk of Jupiter, as the shadow of the moon would be seen to traverse the earth by a spectator favorably situated in space. When Jupiter is east of opposition, the earth being at A (Fig. 65), the shadow strikes the disk of the planet, before the satellite itself is seen upon it, because the satellite will evidently come between S and Jupiter sooner than it does between A and Jupiter. Just the reverse of this takes place, when the earth is at C, that is, when Jupiter is west of opposition. A satellite will then come between C and the planet, sooner than between S and the planet. We do not usually see the satellite itself projected on the disk of the primary, for, being illuminated like the primary, it is not readily distinguishable from it; but sometimes, when it happens to be projected on one of the belts, it is seen as a *bright* spot, making its transit across the disk. Occasionally, also, it is seen as a *dark* spot of smaller dimensions than the shadow. This curious fact has led to the conclusion that certain of the satellites have sometimes on their own bodies, or in their atmospheres, obscure spots of great extent.*

337. A very singular relation subsists between the mean motions of the three first satellites of Jupiter. The mean longitude of the first, plus twice that of the third, minus three times that of the second, always equals 180 degrees. A curious consequence of this relation is, that the three satellites can never be all eclipsed at the same time; for then, having severally the same longitude as the primary, their longitudes would be equal, and that of the first, plus twice that of the third, minus three times that of the second, would be nothing, and of course would not be 180 degrees.† The longitudes here mentioned are such as would be observed by a spectator on Jupiter, and not a spectator on the earth.

338. The discovery of the system of Jupiter and his satellites, soon after the invention of the telescope, lent a powerful support to the Copernican system of astronomy, then just beginning to be received by astronomers, since it presented to the eye an exact miniature of the solar system, and exhibited an actual model of that arrangement of the sun and planets which had before only been contemplated by the eye of the mind; and the laws of the planetary system, discovered by Kepler, were here actually seen to be verified, in the motions of this miniature system. Moreover, the eclipses of Jupiter's satellites, furnished one of those instantaneous events, occurring at the same moment of absolute time wherever seen, which are available for finding the longitudes of different places; and at that period it was deemed a more eligible method of determining this great practical problem of astronomy than any method then in use.

* Sir J. Herschel. † Biot.

339. The eclipses of these satellites seem to have various requisites for determining longitudes, being, as already remarked, seen at the same moment at all places where the planet is visible, being wholly independent of parallax, and being predicted beforehand with great accuracy the instant they occur at Greenwich, and given in the Nautical Almanac: but several circumstances conspire to render this method of finding the longitude less eligible than several other methods at present in use. The extinction of light in the satellite at its immersion, and the recovery of its light at its emersion, are not instantaneous, but gradual; for the satellite, like the moon, occupies some time in entering into the shadow or in emerging from it, which occasions a progressive diminution or increase of light. The better the light afforded by the telescope with which the observation is made, the later the satellite will be seen at its immersion, and the sooner at its emersion.* In noting the eclipses even of the first satellite, the time must be considered as uncertain to the amount of 20 or 30 seconds; and those of the other satellites involve still greater uncertainty. Two observers, in the same room, observing with different telescopes the same eclipse, will frequently disagree in noting its time to the amount of 15 or 20 seconds, and the difference will always be the same way.† Better methods, therefore, of finding the longitude are now employed, although the facility with which the necessary observations can be made, and the little calculation required, still render this method eligible in many cases where extreme accuracy is not required. As a telescope is essential for observing an eclipse of one of these satellites, this method can not be practiced at sea.

340. The grand discovery of the *progressive motion of light* was first made by observations on the eclipses of Jupiter's satellites. In 1675, Roemer, a Danish astronomer, noticed that on comparing the *average* intervals and the *true* intervals between the eclipses of a certain satellite, the true intervals are always shorter when the earth is approaching the planet, so that at the nearest point, the eclipse occurs 16m. $26^s.6$ too

* This is the reason why observers are directed, in the Nautical Almanac, to use telescopes of a *certain power*.

† Woodhouse, p. 810.

soon. And, as the earth departs from Jupiter, Roemer perceived that the eclipses occur later than they should do according to the average interval, and that they are 16m. 26s.6 too late, when at the greatest distance. He attributed this effect to the progressive motion of light. Dividing 190,000,000 miles, the diameter of the earth's orbit, by 16m. 26s.6, the time of crossing, he found 192,600 miles per second to be the velocity of light. This seemed, at first, quite incredible, and was received with distrust. But its correctness was soon established, by the discovery of the aberration of the stars, which gives the same result.

341. SATURN comes next in the series as we recede from the sun, and has, like Jupiter, a system within itself, on a scale of great magnificence. In size it is, next to Jupiter, the largest of the planets, being 79,000 miles in diameter, or nearly 10 times as large as the earth in diameter, and about 1000 times as large in volume. It has likewise belts on its surface, and is attended by eight satellites. But a still more wonderful appendage is its *Ring*, a broad wheel encompassing the planet at a great distance from it. We have already intimated that Saturn's system is on a grand scale. As, however, Saturn is distant from us nearly 900,000,000 miles, we are unable to obtain the same clear and striking views of his phenomena that we do of the phenomena of Jupiter, although they really present a more wonderful mechanism. The disk of Saturn was described by Sir William Herschel as having the form of a rectangle with rounded corners; but refined measurements, more recently made, show that it is an ellipse, and the planet, therefore, an ellipsoid. Its equatorial exceeds its polar diameter by about one-tenth.*

The *belts* of Saturn, although clearly discerned by a good telescope, are far more indistinct than those of Jupiter. Spots, which occasionally appear on the belts, have enabled astronomers to determine the time of the diurnal rotation of Saturn, which is found to be about ten hours and a half (10h. 29m.).

342. When viewed with a good telescope, the body of the

* Hind.

planet is seen to be surrounded by a broad thin ring, placed obliquely to our line of vision, so that it appears elliptical, showing one part in front of the planet and having a part hid behind it. At the ends of the ellipse, and sometimes entirely around it, is seen a division of the ring into two parts, of which the inner is the broadest. Though this division, on account of its immense distance, appears as a delicate line, yet it is, in reality, an interval of 1800 miles. The dimensions of the whole system, in round numbers, are as follows:*

	Miles.
Diameter of the planet, ...	79,000
From the surface of the planet to the inner ring,	20,000
Breadth of the inner ring, ...	17,000
Interval between the rings, ...	1,800
Breadth of the outer ring, ...	10,500
Extreme dimensions from outside to outside, .	176,000

The figure (Frontispiece) represents Saturn, as it appears to a powerful telescope, surrounded by its rings, and having its body striped with dark belts, somewhat similar, but broader and less strongly marked than those of Jupiter, and owing, doubtless, to a similar cause.† The ring is opaque, since it casts a deep shadow on the planet. The earth is generally so situated that we can see some part, both of the shadow cast by the ring on the planet, and by the planet on the ring.

The rings of Saturn have been long and carefully observed by Prof. G. P. Bond, with the celebrated Cambridge refractor. He finds that there is a third ring within the others, which reflects a dim light, and is about half as wide as the inner bright ring, but not separated from it by a noticeable interval. He also finds the two bright rings to be divided at times into several rings, with very narrow intervals. In one instance the outer ring was divided into two, in another, into four parts; and the inner one, in some cases, into six or eight delicate rings. The dark separating lines are usually traceable only at the ends of the ellipse. The investigations of Professor Peirce, aided by Bond's observations, have nearly, if not fully, established the fact of the fluidity of the rings, or at least, of the

* Professor Struve, Mem. Ast. Soc., iii., p. 301.

† Sir J. Herschel.

incoherency of their parts, if they consist of solid matter. A changeableness of form and condition is indicated, also, when the ring presents to us its edge view. It sometimes appears as a delicate, uniform line; at others, as a line of unequal thickness, whose thinner parts occasionally become entirely invisible; and the inequalities change their aspects, from time to time, in a manner not to be accounted for by the revolution which the system is known to have.

A most remarkable fact relating to the rings, is their exceeding thinness. They have generally been regarded as about 100 miles thick. But Bond's observations lead to the conclusion that their thickness is less than 40 miles. If a model of the rings, one foot in diameter, were cut out of common writing-paper, the thickness would be too great to represent them properly.

343. *Saturn's ring, in its revolution around the sun, always remains parallel to itself.*

If we hold, opposite to the eye, a circular ring or disk, like a piece of coin, it will appear as a complete circle when it is at right angles to the axis of vision; but when oblique to that axis, it will be projected into an ellipse more and more narrow, as its obliquity is increased, until, when its plane coincides with the axis of vision, it is projected into a straight line. Let us place on the table a lamp or a ball, to represent the sun, and, holding the ring at a certain distance, inclined a little toward the central body, let us carry it round, always keeping it parallel to itself. During its revolution it will twice present its edge to the lamp or ball at opposite points, and twice, at 90° distance from those points, it will present its broadest face toward the central body. At intermediate points it will exhibit an ellipse more or less open, according as it is nearer one or the other of the preceding positions. It will be seen, also, that in one-half of the revolution the lamp shines on one side of the ring, and in the other half of the revolution on the other side. Such would be the successive appearances of Saturn's ring to a spectator on the sun; and since the earth is, in respect to so distant a body as Saturn, very near the sun, those appearances are presented to us nearly in the same manner as though we viewed them from the sun. Accordingly, we sometimes see Saturn's ring under the form of a broad ellipse, which grows continually more and more narrow until it passes into a line, and we either lose sight of it altogether, or, with the aid of the most powerful telescopes, we see it as a fine line drawn across the disk, and projecting out from it on each side. As the whole revolution occupies nearly 30 years, and the plane of the ring passes across the earth's orbit twice in the revolution, the phenomena attending the edge view occur every 15 years, when the ring may within a single year disappear repeatedly, and for different reasons, as described in Art. 346.

344. The learner may perhaps gain a clearer idea of the foregoing appearances from the following diagram.

Let A, B, C, &c., Fig. 66, represent successive positions of Saturn and his ring in different parts of his orbit, while *ab*

Fig. 66.

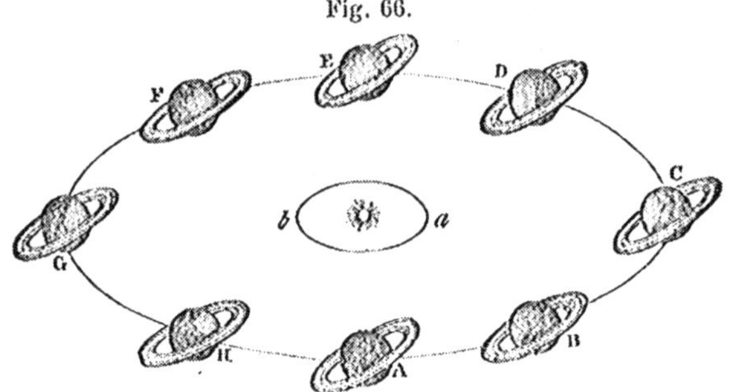

represents the orbit of the earth.* Were the ring when at C and G perpendicular to the line joining CG, it would be seen by a spectator situated at *a* or *b* as a perfect circle, but being inclined to the line of vision 28° 11′, it is projected into an ellipse. This ellipse gradually contracts to a straight line, as the ring passes to A and E, where its nodes are in the direction of the earth's orbit. From E to G the ellipse widens again, till at G the breadth is nearly half as great as its length. Through the quadrant from G to A it contracts, and from A to C it expands, occupying about 7½ years in going from the

* It may be remarked by the learner, that these orbits are made so elliptical, not to represent the *eccentricity* of either the earth's or Saturn's orbit, but merely as the projection of circles seen very obliquely.

maximum breadth to the minimum, and the reverse. These changes are visible in a telescope of moderate powers, except that the ring is too thin to be seen at A and E.

345. There are three ways in which the ring may fail to be visible during the period when the line of its nodes is crossing the earth's orbit. 1st. It may present its edge exactly to the *earth*, when in common telescopes, it subtends too small an angle to be seen. 2d. It may present its edge exactly to the *sun*, so that neither side of the ring is enlightened. 3d. Its plane may be directed *between* the earth and sun, when the dark side is toward us. The two first causes may be considered as only momentary; for the plane of the ring passes the breadth of the sun in less than two days, and of the earth in about 20 minutes. But the third cause may continue to render the ring invisible for several months.

In Fig. 67 let S be the sun, C the place of Saturn, when the

Fig. 67

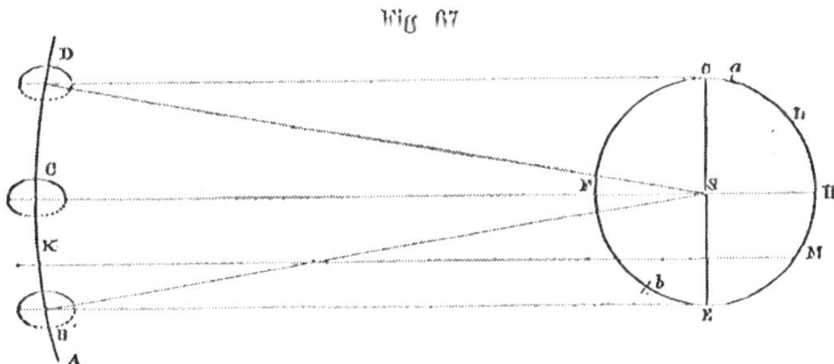

nodes of its ring are in the line CS, passing through the sun. Let EFGH be the earth's orbit; then EB, GD, parallel to SC, will include BD, the arc of Saturn's orbit in which the ring will present its edge to the earth's orbit. If the orbits are supposed to be in the plane of the paper, we must conceive the small ellipses at B, C, D, representing the ring, to be inclined about 28°, having their section with the orbit, in BE, CS, and DG, respectively. While Saturn passes over the arc BD, the plane of its ring will pass once through the sun, and may repeatedly pass through the earth, on account of the revolution of the latter. The time of describing BD differs only about six days from a year; for, since Saturn is 9.54 times as far from the sun as the earth is, therefore, 9.54 : 1 : : SB : SE : : rad : sin SBE; which is thus found to be 6° 1′. Hence, BSC =6° 1′, and BSD=12° 2′. But Saturn describes 12° 2′ in 359½ days, near six days less than a year. The earth therefore very nearly describes the orbit EFGH, while Saturn describes the arc BCD.

346. The disappearances of the ring may be variously modified during the year of passing the node, according to the place of the earth when the nodal line first reaches its orbit. A few cases are here described. If the earth is at G, when the plane of the ring passes E, then, while the nodal line moves from BE to CS, the earth will go from G through H to E, and must cross that line below H. The ring will *begin* its disappearance at that moment; but the disappearance will *continue*, as the earth proceeds to E, because the dark side is toward it. This disappearance will last about two months, and *close* when the plane of the ring at C passes the sun; for after that, the illuminated side will be toward the earth. While the earth proceeds from E through F to G, the plane of the ring will move from CS to DG, before the earth can overtake it.

Again, if the earth had advanced some distance on the quadrant GH, when the ring's line of nodes was at E, then they will meet further from H than before, say at M; after which the dark side will be toward the earth. The plane of the ring will pass the sun, when the earth is on the quadrant EF, after which the bright side is presented to the earth. But the earth will overtake the nodal line before reaching G, and therefore look again upon the dark side, until it recrosses the line on the quadrant GH. Thus there are two periods of disappearance. These two may possibly unite in one, of 8 months' continuance; this will happen when the earth and node-line pass F at the same time; for then the plane of the ring is between the earth and sun both before and after passing through the sun.

It is a possible thing, that no disappearance at all should happen during the nodal year. Suppose the earth at F, when the line of nodes arrives at E. Then, while the line moves from BE to CS, the earth will describe FGH, all the time on the luminous side of the ring; the earth and sun will both be in the line CS at once, the planet being in conjunction with the sun; and, after the earth has passed H toward F, the bright side of the ring is again in view. Thus, there is only a momentary disappearance, and that when the planet itself is lost in the blaze of the sun's light.

But in general there are two periods of disappearance within the nodal year, arising from the *third* cause (Art. 345), each beginning and ending with a disappearance from the *first* or *second* cause.

347. The rings of Saturn must present a magnificent spectacle to that hemisphere of the planet to which their illuminated side is turned, appearing as bright arches several degrees in width, and spanning the sky from one side of the horizon to the other. They revolve diurnally in the same plane as the planet itself, and in about the same time, $10\frac{1}{2}$ hours.

348. Saturn is attended by *eight* satellites. Their sizes vary from 500 to 2,850 miles;* but, on account of their great distance, they are seen only with the best instruments. They are all external to the rings, and the eighth at the distance of 2,500,000 miles from the planet.† The seventh was discovered in Cambridge, Mass., by Professor Bond. Their orbits are nearly in the plane of the ring, and make an angle of about 28° with the orbit of the planet. Only the two interior ones are eclipsed, except when the ring is seen edgewise.‡

349. Uranus, the next planet in the series, was discovered by Sir William Herschel, in 1781. Previous to this time, Saturn had, from a high antiquity, been considered as the outermost boundary of the solar system; but this discovery doubled the dimensions of the system, bringing to light a large planet at about twice the distance of Saturn from the sun, and about 19 times the distance of the earth, or 1800 millions of miles. It was named by the discoverer the *Georgian*, in honor of his patron, George III.; but this name being unacceptable

* Hind.

† They have received the names, Mimas, Enceladus, Tethys, Dione, Rhea, Titan, Hyperion, and Japetus.

‡ Sir J. Herschel.

to astronomers of other countries, the planet was called *Herschel* in America, after the name of the discoverer, and *Uranus** on the continent of Europe, which last appellation is now universally adopted. The diameter of Uranus is about 35,000 miles, and consequently its volume more than 80 times that of the earth. Its revolution around the sun occupies nearly 84 years, so that its position among the stars varies but little for several years in succession, since it shifts its place only a little more than four degrees in a year, and of course would remain in the same sign of the Zodiac seven years. Its path lies very near the ecliptic, being inclined to it less than 0° 47′. The sun himself, when seen from Uranus, dwindles almost to a star, subtending, as it does, an angle of only 1′ 40″; so that the surface of the sun would appear there nearly 400 times less than it does to us.

350. The *satellites* of Uranus are exceedingly minute objects, and visible only to the most powerful telescopes. Although Sir William Herschel assigned *six* satellites to this planet, yet only two of the number (the second and fourth in the order of distances) have, until quite recently, been seen by other astronomers. Two others have of late been added, and an increasing confidence is beginning to be felt that the entire number announced by Herschel will be identified. The orbits of these satellites, says Sir John Herschel, offer remarkable, and indeed quite unexpected and unexampled peculiarities. Contrary to the unbroken analogy of the whole planetary system, whether of primaries or secondaries, the planes of their orbits are nearly *perpendicular to the ecliptic*, being inclined no less than 78° 58′ to that plane, and in these orbits their motions are retrograde. Instead of advancing from west to east, as is the case with every other planet and satellite, they move in the opposite direction, or from east to west. With this exception, all the motions of the planets, whether around their own axes or around the sun, and that of the sun himself on his axis, are from west to east.

351. NEPTUNE is (so far as is known) the last planet of the

* From ουρανος, the father of Saturn.

series, being removed from the sun to the immense distance of nearly 3000 millions of miles (2,862,457,000). Its diameter is a little less than that of Uranus, being 31,000 miles.* Its volume is nearly sixty times that of the earth. Its periodic time is 164½ years, which is about twice that of Uranus. Its orbit is nearly circular, and but little inclined to the ecliptic (1° 47′).

The discovery of the planet Neptune is the most remarkable astronomical event of our times, and is generally considered as the most extraordinary discovery ever made in physical science. The leading steps of the process were as follows. The planet Uranus had long been known to be subject to certain irregularities in its revolution around the sun, not accounted for by all the known causes of perturbation. In some cases the deviation from the true place, as given by the tables, differs from actual observation two minutes of a degree—a quantity indeed which seems small, but which is still far greater than occurs in the case of the other planets, and far too great to satisfy the extreme accuracy required by modern astronomy. This fact long since suggested to astronomers the possibility of one or more additional planets, hitherto undiscovered, which, by their attractions, exert on Uranus a great disturbing influence. Le Verrier, a distinguished French astronomer, assuming the existence of such a planet, applied himself, by the aid of the calculus, guided by the law of universal gravitation, to the inquiry *where* the hidden planet was situated—at what distance from the sun—and at what point of the starry heavens? From Bode's law of the planetary distances (Art. 299), according to which Saturn is nearly twice as far from the sun as Jupiter, and Uranus twice as far as Saturn, he inferred that, if a planet exists beyond Uranus, its distance is probably about twice that of Uranus, or about 3,600 millions of miles from the sun, which is nearly thirty-eight times that of the earth. He assumed it, however, to be thirty-six times the earth's mean distance. The corresponding periodic time would be 216 years. After reasoning from analogy, and the doctrine of universal gravitation, respecting the position and mass which a body must have in order to account for the perturbations of Uranus, equations were formed between these perturbations and the

* Hind.

assumed and unassumed elements of the body in question. These equations were exceedingly complex and difficult of reduction; but, by the most ingenious artifices, the several unknown quantities were successively eliminated, either directly or by repeated approximations, until the great geometer arrived at expressions for the elements of the unknown planet, which indicated its place among the stars, its quantity of matter, the shape of its orbit, and the period of its revolution. Having placed the body in various positions in the orbit thus determined, he found that when situated at a point in the constellation Capricornus, its effect upon Uranus would be such as corresponded to the irregularities to be accounted for; that on the 1st of January, 1847, the hidden planet would have a longitude of 326° 32′, and would lie about five degrees eastward of the well-known star *Delta Capricorni.* He further asserted that it would have an apparent diameter of about 3″, and therefore be visible to large telescopes.

352. Having communicated these results to the French Academy, at their sitting on the 31st of August, 1846, Le Verrier soon afterward made them known to Dr. Galle, one of the astronomers of the Royal Observatory of Berlin, with the request that he would search for the stranger with the powerful telescope at his command. On the same evening that Dr. Galle received the communication, namely, on the 23d of September, he directed his telescope toward the spot assigned for the planet, and there it was, within less than a degree of the place indicated by Le Verrier, and having an apparent magnitude within half a second of that assigned. To show the near correspondence between theory and observation, we may remark that the predicted longitude, for the 23d of September, at midnight, was 324° 58′, and the observed longitude was 325° 52′.8; the predicted diurnal motion in longitude was 69″, and the observed 74″. These results struck the scientific world with astonishment, and their confirmation was one of the greatest achievements of the human mind.

353. It has often happened, in the history of great discoveries, that the same hidden truth is revealed simultaneously to different inquirers, and accordingly, by a singular coincidence, a young mathematician of the University of Cambridge (Eng.), Mr. Adams, had, without the least knowledge of what M. Le Verrier was doing, arrived at the same great result. But having failed to publish his paper until the world was made acquainted with the facts through the other medium, he has lost much of the honor which the priority of discovery would have gained for him. Thus two distinguished mathematicians, unknown to each other, and by entirely independent processes, had arrived at the same results, as regarded both the existence of the supposed planet, and the region of the starry heavens where at that moment it lay concealed; and, to crown all, astronomers, in obedience to the direction of one of them, had pointed their telescopes to the spot and found it there. The conviction on the mind of every one was, that nothing but absolute truth could abide a test so unequivocal. It still remained, however, to determine by observation whether the body actually conformed, in *all* respects, to the results of theory. To settle this point completely, that is, to determine with precision the elements of the orbit from observation, would require a long time in a planetary body whose motion was so slow that more than two centuries, as was supposed, would be required to complete a single revolution. But if it should be found that, among preceding catalogues of the stars, this body might have been included, and its place recorded as a fixed star, then, by comparing that place with its present position, and noting the interval of time between the two observations, we might thus learn the rate of its motion, and its periodic time, and might thence deduce various other particulars dependent on these elements. Our distinguished countryman, Mr. Sears C. Walker, then connected with the observatory at Washington, undertook this investigation. First, from the observations already accumulated, he calculated the path which the planet must have pursued for the last fifty or sixty years, and by tracing this path among the stars of Lalande's catalogue, he found that it passed within two minutes of a star of the seventh magnitude, which was recorded as being seen in May, 1795. Professor Hubbard, of the same observatory, on reconnoitering for this star, found that it was missing. Little doubt remained that the star seen by Lalande was the planet of Le Verrier; and this conclusion was con firmed by calculating its orbit on this supposition, and comparing the results with the places it has actually occupied since it fell within the sphere of observation. The results thus obtained, however, were materially different from those of Le Verrier and Adams. Instead of a period of 216 years, they give only a period of 164½ years; and instead of a distance of 3,600 millions of miles, the new period would require a distance of only 2,862 millions. The eccentricity of the orbit, moreover, according to Walker, is much less than had been assigned to it, the orbit being in fact very nearly circular, while, by Le Verrier's estimate, it was considerably elliptical. The longitude, in fact, proved to be nearly the same as that assigned to it. But this close agreement is to be considered accidental; for Le Verrier himself fixed on the precise place which he named, as only the most *probable*. On account of uncertainty in the data, he stated that there might be a variation of 9° either way.* The elements thus corrected account fully and completely for the irregularities of Uranus sought to be explained, within a single second, as determined by Professor Peirce.†

MOTIONS OF THE PLANETARY SYSTEM.

354. We have waited until the learner may be supposed to be familiar with the heavenly bodies individually, before inviting his attention to a systematic view of the planets in their revolutions around the sun, and their grand laws.

There are two methods of arriving at a knowledge of the motions of the heavenly bodies. One is, to begin with the *apparent*, and from these to deduce the *real* motions; the other is to begin with considering things as they really are in nature, and then to inquire why they appear as they do. The latter of these methods is by far the more eligible. It is much easier than the other; and proceeding from the less difficult to that which is more so—from motions which are very simple to such as are complicated, it finally puts the learner in possession of the whole machinery of the heavens. We shall in the first place, therefore, endeavor to introduce the student to an ac-

---

* Loomis's Recent Prog. of Astron., pp. 50–52.

† Amer. Journal of Science, *New Series*, vol. v., p. 436.

quaintance with the simplest motions of the planetary system, and afterward to conduct him gradually through such as are more complicated and difficult.

355. When viewed from the center of their motions, the revolutions of the planets would appear simple and harmonious, all coursing round the spectator from west to east in regular order, in nearly the same great highway, though with very different degrees of velocity. Let us, then, suppose ourselves standing on the sun, and contemplate the revolutions of the planets, first, severally, and then as forming one grand whole, consisting of numerous parts, but bound together under the same laws in one vast empire. We should see Mercury making very perceptible progress around the heavens, like the moon in its motions about the earth, his rate of motion eastward being about one-third as rapid as that of the moon, since he completes his entire revolution in about three months. It will, at first, aid our conceptions of the respective positions of the planetary orbits, to imagine the ecliptic to be marked out on the face of the visible heavens in a palpable line distinctly visible to the eye. If we, stationed at the sun, watch the motions of Mercury, we shall see it cross the ecliptic in two opposite points of the heavens, constituting its *nodes;* and we shall see it, when half way between the nodes, at an angular distance from the ecliptic of about 7°, this being the *inclination* of its orbit. Knowing the position of the orbit of Mercury with respect to the ecliptic, we may now, in imagination, represent that orbit in a great circle passing through the center of the planet and the center of the sun, and cutting the plane of the ecliptic in two opposite points in an angle of 7 degrees. The paths of both planets appear as circles among the stars; and if we suppose them to be visibly traced on their respective planes, the observer, while at the sun, can not distinguish them from circles, of which he is the center. But, if we imagine him transported to a great distance from the sun, in a line at right angles to the ecliptic, he will discern the true forms of the orbits. The earth's orbit can not even now be distinguished from a circle, but the sun is plainly a little out of its center. The orbit of Mercury, however, is distinctly elliptical, with the sun in one of its foci. On his return to his station at the sun, he loses all perception of this elliptical form, and of his eccentric position. But he perceives the *consequences* of these facts,—an alternate increase and decrease of size and of velocity in the planets. The earth slowly becomes larger, and moves more swiftly, till it reaches a certain position, and then diminishes both in size and velocity, and attains a minimum at a point 180° from the maximum. Mercury passes through still greater changes of the same kind.

356. A clear understanding of the motions of Mercury, and of the relations of its orbit to the plane of the ecliptic, will render it easy to understand the same particulars in regard to each of the other planets. Standing on the sun, we should see each of the planets pursuing a similar course to that of Mercury, all moving from west to east, differing from each other chiefly in two respects, namely, in their velocities, and in the distances to which they recede from the ecliptic, or their inclinations. We have supposed the observer to select the plane of the earth's orbit as his standard of reference, and to see how each of the other orbits is related to it; but such a selection of the ecliptic is entirely arbitrary; the spectator on the sun, who views the motions of the planets as they actually exist in nature, would make no distinction between the different orbits, but merely inquire how they are mutually related to each other. Taking, however, the ecliptic as the plane to which all the others are referred, we do not, as in the case of the other planets, inquire how its plane is *inclined*, nor what are its *nodes*, since it has neither inclination nor node.

357. Such, in general, are the *real* motions of the planets, and such the appearances which the planetary system would exhibit to a spectator at the center of motion. But, in order to represent correctly the positions of the planetary orbits, at any given time, *three* things must be regarded: the *Inclination* of the orbit to the ecliptic; the position of the *line of the Nodes*, and the position of the *line of the Apsides.* In our common diagrams, the orbits are incorrectly represented, being all in the same plane, as in the following diagram, where AEB (Fig. 68) represents the orbit of Mercury as lying in the same plane with the ecliptic. To exhibit its position justly,

AB being taken as the line of the nodes, the plane should be elevated on one side about 7°, and depressed the same number of degrees on the other side, turning on the line AB as on a hinge. But even then the representation may be incorrect in other respects, for we have taken it for granted that the line of the nodes coincides with the line of the apsides, or that the orbit of Mercury cuts the ecliptic in the line AB, the major axis of the orbit, whereas it may lie in any given position with respect to the line of apsides, according to the longitude of the nodes. If, for example, the line of nodes had chanced to pass

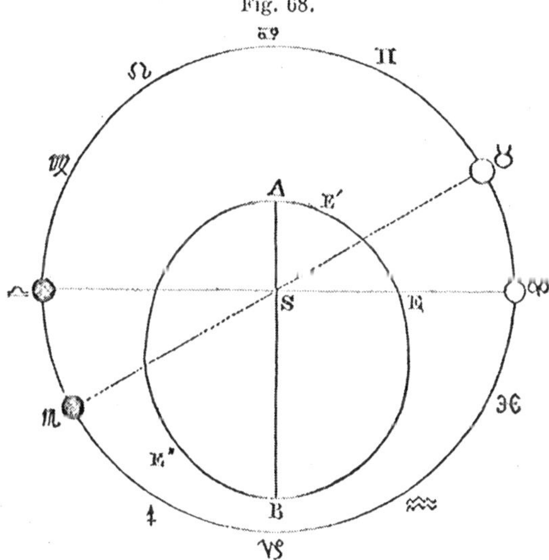

Fig. 68.

through Taurus and Scorpio instead of Cancer and Capricorn, then it would have been represented by the line ♉♏ instead of the line passing through ♋, and the plane, when elevated or depressed with respect to the plane of the ecliptic, would be turned on this line in our figure. Moreover, our diagram represents the line of apsides as passing through Cancer and Capricorn, whereas it may have any other position among the signs, according to the longitudes of the perigee and apogee.

358. Having acquired as correct an idea as we are able of the planetary system, as seen from the sun, and of the positions of the orbits with respect to the ecliptic, let us next inquire into the nature and causes of the *apparent motions.* The apparent motions of the planets are exceedingly unlike the real motions, a fact which is owing to two causes: first, we view them out of the center of their orbits; secondly, *we are ourselves in motion.* From the first cause, the apparent places of the planets are greatly changed by perspective; and, from the second cause, we attribute to the planets changes of place which arise from our own motions, of which we are unconscious.

359. The situation of a heavenly body, as seen from the center of the sun, is called its *heliocentric* place; as seen from the center of the earth, its *geocentric* place. The geocentric motions of the planets must, according to what has just been said, be far more irregular and complicated than the heliocentric, as will be evident from the following diagram, which represents the geocentric motions of Mercury for two entire revolutions, embracing a period of nearly six months. Let S (Fig. 69) represent the sun, 1, 2, 3, &c., the orbit of Mercury, *a*, *b*, *c*, &c., that of the earth, and GT the concave sphere of the heavens. The orbit of Mercury is divided into 12 equal parts, each of which he describes in $7\frac{1}{3}$ days; and a portion of the earth's orbit described by that body in the time that Mercury describes the two complete revolutions, is divided into 24 equal parts. Let us now suppose that Mercury is at the point 1 in his orbit, when the earth is at the point *a*; Mercury will then appear in the heavens at A. In $7\frac{1}{3}$ days Mercury will have reached 2, while the earth has reached *b*, when Mercury will appear at B. By laying a ruler on the point *c* and 3, *d* and 4, and so on, in the order of the alphabet, the successive apparent places of Mercury in the heavens will be obtained. From A to C, the apparent motion is direct, or in the order of the signs; from C to G it is retrograde; at G it is stationary a while, and then direct through the whole arc GT. At T the planet is again stationary, and afterward retrograde along the arc TX. Hence it appears that the motions of an inferior planet, as viewed from the earth, are exceedingly irregular and complicated, although it is all the while pursuing its course at a nearly uniform rate, and in the same unvarying direction around the sun. It moves forward when near the superior conjunction, backward when near the inferior, and is stationary near the points of greatest elongation. The planet moves sometimes very slowly, and then rapidly; at one time backward over a small space, and then forward for a great distance.

Fig. 69.

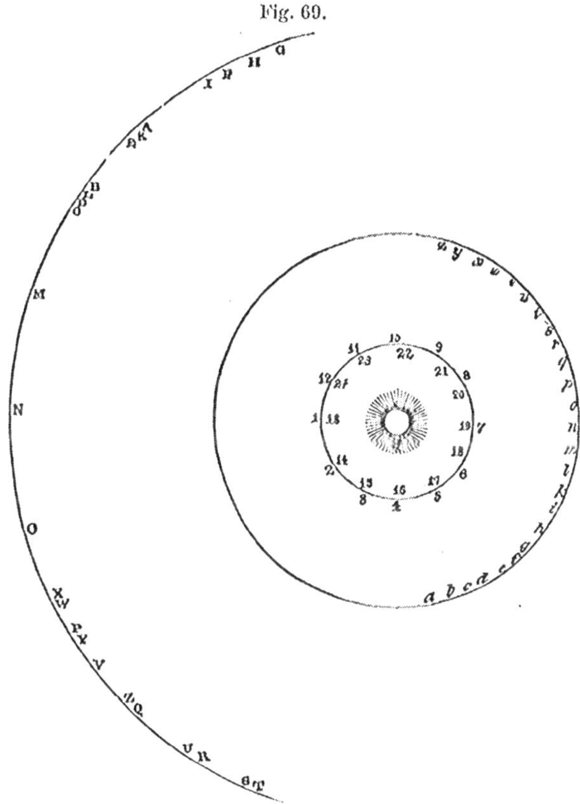

Yet all these apparent irregularities are owing to the two causes already adverted to, viz., the effects produced by perspective, and by the motions of the spectator himself. Venus exhibits a variety of motions similar to those of Mercury, except that the changes do not succeed each other so rapidly, since her period of revolution approaches more nearly to that of the earth.

360. The apparent motions of the *superior* planets are, like those of Mercury and Venus, alternately direct and retrograde, and between the two the planets are stationary. In this case, however, the earth moves faster than the planet, and the planet has its opposition, but no inferior conjunction; whereas an inferior planet has its inferior conjunction, but no opposition. These differences render the apparent motions of the superior planets in some respects unlike those of Mercury and Venus. On the side of the sun most remote from the earth, the motion of a superior planet is direct, because, as is the case with Venus in her superior conjunction (see figure 61), the only effect of the earth's motion is to accelerate it; but when the planet is in opposition, the earth is moving past it with greater velocity, and makes the planet seem to move backward, like the apparent backward motion of a vessel when we overtake it and pass rapidly by it in a steamboat.

361. Let ABCD (Fig. 70) represent the earth in different positions in its orbit, M a superior planet as Mars, and NR an

Fig. 70.

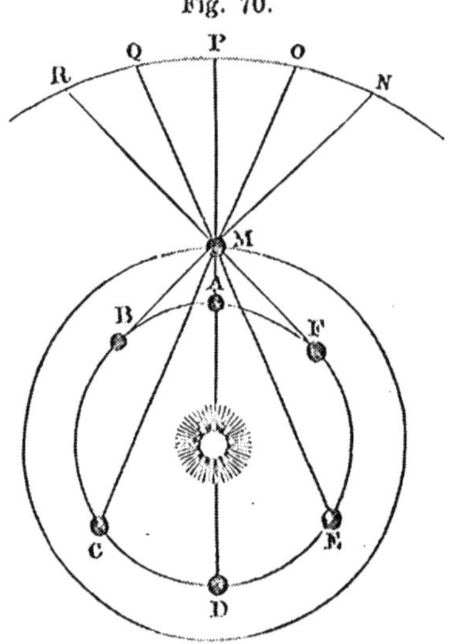

arc of the concave sphere of the heavens. First, suppose the planet to remain at rest in M, and let us see what apparent motions it would receive from the real motions of the earth. When the earth is at B, it will see the planet in the heavens at N; and as the earth moves successively through CDEF, the planet will appear to move through OPQR; B and F are the two points of greatest elongation of the earth from the sun, as seen from the planet: between these two points, while passing through the part of its orbit most remote from the planet (at which time the planet is seen in superior conjunction), the earth, by its own motion, gives an apparent motion to the planet in the order of the signs; that is, the apparent motion given by the earth's motion, when the planet is seen toward its superior conjunction, is *direct.* But in passing from F to B through A, when the planet is seen toward its opposition, the apparent motion given to the planet by the earth's motion is *retrograde.* But the superior planets are not in fact at rest, as we have supposed, but are all the while moving eastward, though with a slower motion than the earth. Indeed, with respect to the remotest planets, as Saturn and Uranus, the forward motion is so exceedingly slow, that each remains for a long time in the same sign of the zodiac. Still, the effect of the real motions of all the superior planets eastward, is to increase the direct apparent motion communicated by the earth, and to diminish the retrograde motion, as will be readily seen from the figure.

CHAPTER XI.

DETERMINATION OF THE PLANETARY ORBITS—KEPLER'S DISCOVERIES—ELEMENTS OF THE ORBIT OF A PLANET—QUANTITY OF MATTER IN THE SUN AND PLANETS.

362. In Chapter II. we have shown that the figure of the *earth's* orbit is an ellipse, having the sun in one of the foci, and that the earth's radius describes equal spaces in equal times; and in Chapter III. we have remarked that these are only particular examples under the law of Universal Gravitation, as is also the additional fact, that the squares of the periodic times of the planets are as the cubes of the major axes of their orbits. We may now learn more particularly the *process* by which the illustrious Kepler was conducted to the discovery of these grand laws of the planetary system. From the apparent motions of the heavenly bodies, as seen projected on the face of the sky, the ancient astronomers inferred that their orbits were necessarily circular, and the motions actually uniform. Still, Hipparchus and Ptolemy were not ignorant of the fact that the sun moves faster through the winter than through the summer signs, performing the half of his revolution around the earth nearly eight days sooner from the autumnal to the vernal, than from the vernal to the autumnal equinox. This led them to infer that the earth is not in the center of the circle, but nearer to one side of the circle than to the other, by which means the sun would appear to move more rapidly in that part of its orbit than in the opposite part, just as a steamboat appears, to a spectator on the shore, to move faster when nearer than when more remote from the shore, although her actual speed is the same in both cases. On a similar supposition Tycho Brahe made a great number of very accurate *observations* on the planetary motions, which served Kepler as standards of comparison for results, which he deduced from *calculations*, founded on the application of geometrical reasoning to various hypotheses which he successively assumed as to the figure of the planetary orbits; first supposing the orbit to be of a certain figure, then determining from the geometrical properties of the curve what motions the body would appear to us to have when moving in such a path, and finally testing his conclusions by comparing them with the facts, as determined by Tycho, from observation.

363. Kepler first applied himself to investigate the figure of the orbit of Mars, the motions of which planet appeared more irregular than those of any other planet except Mercury, which, being seldom seen, had been very little studied. Like Ptolemy and Tycho, he first supposed the orbit to be circular, and the planet to move uniformly about a point at a certain distance from the sun. He made *seventy* suppositions before he obtained one that agreed with observation, the calculation of which was extremely long and tedious, occupying him more than five years.* The supposition of an equable motion in a circle, however varied, could not be made to conform to the observations of Tycho, whereas the supposition that the orbit was an oval figure, depressed at the sides, but coinciding with a circle at the perihelion, agreed so nearly with observation as to leave no doubt that the orbit of Mars is an ellipse, having the sun in one of its foci. He immediately inferred that the same is true of the orbits of all the other planets; and a similar comparison of this hypothesis with observation, confirmed its truth. Thus he established the first great law, viz., *The planets revolve about the sun in ellipses, having the sun in one of the foci.*

364. Kepler also discovered, from observation, that the velocities of a planet, when in the apsides of its orbit, are inversely as the distances, and therefore the product of the velocity into the distance would, in those two points, make the same quantity. But the velocity is the length of arc described in a unit of time; and the length of an arc multiplied by its radius, is double the sector upon that arc. Therefore the area described by the radius vector at one apsis equals that described at the other apsis in the same time. (Fig. 32, p. 86.) Although he could not prove, from *observation*, that the same was true in every point of the orbit, yet analogy suggested that such was probably the fact. Therefore, assuming this principle as true, and hence deducing the equation of the center (Art. 200), he found the result to agree with observation, and thus arrived at the conclusion (which has since been proved true (Art. 181) from the principles of common mechanics), that *the radius vectors of the planetary orbits describe about the sun equal areas in equal times.*

365. Having in his researches, that led to the discovery of the first of the above laws, found the *relative* mean distances of the planets from the sun (Art. 308), and knowing their

* Logarithms were invented during the age of Kepler, but were not available to him until his most laborious calculations had been performed. In relation to these, he expresses himself thus: *Si te hujus laboriosæ methodi pertæsum fuerit, jure mei te misereat, qui eam ad minimum septuagies ivi cum plurima temporis jactura; et mirari desines hunc quintum jam annum abire, ex quo Martem aggressus sum.*

periodic times from observation, Kepler next endeavored to ascertain if there was any relation between the distances and times of revolution, having a strong passion for tracing analogies in nature. He saw at once that the more distant a planet is from the sun, the slower it moves; so that the periodic times of the remoter planets are increased on two accounts: first, because they have a longer path to traverse; and secondly, because they actually move more slowly in their orbits than the planets nearer the sun. Saturn, for example, is $9\frac{1}{2}$ times further from the sun than the earth is; and since the circumferences of circles are as their radii, the orbit of Saturn must be larger than the earth's in the same ratio; so that if the periodic time of Saturn were longer than the earth's, merely because its orbit is larger, that period would be $9\frac{1}{2}$ years, whereas it is 30 years. Hence it is evident that the periodic times of the planets increase in a greater ratio than their distances from the sun, but in a less ratio than the squares of the distances, for then the time of Saturn would be about 90 years. Kepler then compared the squares of the times with the cubes of the distances, and found an exact agreement between them. Thus he discovered the famous law, *the squares of the periodic times of all the planets are as the cubes of their mean distances from the sun.**

366. This law is strictly true only in relation to planets whose quantity of matter in comparison with that of the central body is inappreciable. When this is not the case, the periodic time is shortened in the ratio of the square root of the sun's mass divided by the sun's plus the planet's mass, as expressed by the formula $\left(\frac{M}{M+m}\right)^{\frac{1}{2}}$. The mass of the planets is, however, so small compared to the sun's, that this modification of the law is unnecessary except where extreme accuracy is required.

ELEMENTS OF THE PLANETARY ORBITS.

367. The particulars necessary to be known in order to determine the precise situation of a planet at any instant, are

* Vince's Complete System, i., p. 98.

called the *Elements of its Orbit.* They are seven in number: of which the first two determine the position of the plane of the orbit, and the other five relate to the orbit and the planet in that plane. These elements are:

1. *The position of the line of the nodes.*
2. *The inclination to the ecliptic.*
3. *The periodic time.*
4. *The mean distance from the sun, or semi-axis major.*
5. *The eccentricity.*
6. *The place of the perihelion.*
7. *The place of the planet in its orbit at a particular epoch.*

368. It may at first view be supposed that we can proceed to find the elements of the orbit of a planet in the same manner as we did those of the solar or lunar orbit, namely, by observations on the right ascension and declination of the body, converted into latitudes and longitudes by means of spherical trigonometry (see Art. 132). But in the case of the moon, we are situated in the center of her motions, and the apparent coincide with the real motions: and in respect to the sun, our observations on his *apparent* motions give us the earth's *real* motions, allowing 180° difference in longitude. But as we have already seen, the motions of the planets appear exceedingly different to us, from what they would if seen from the center of their motions. It is necessary, therefore, to deduce from observations made on the earth the corresponding results as they would be if viewed from the center of the sun; that is, in the language of astronomers, having the *geocentric* place of a planet, it is required to find its *heliocentric* place.

369. The first steps in this process are the same as in the case of the sun and moon. That is, for the purpose of finding the right ascension and declination, the planet is observed on the meridian with the Transit Instrument and Mural Circle (see Arts. 155 and 230), and from these observations, the planet's *geocentric* longitude and latitude are computed by spherical trigonometry. The distance of the planet from the sun is known nearly by Kepler's law. From these data it is required to find the heliocentric longitude and latitude.

Let S and E (Fig. 71) be the sun and earth, ASOEH the

plane of the ecliptic, SA and EH parallels in that plane, pointing to the vernal equinox (which is considered infinitely distant), P the planet, PO the perpendicular from it to the plane of the ecliptic. Let HEO, the angular distance in the plane

Fig. 71.

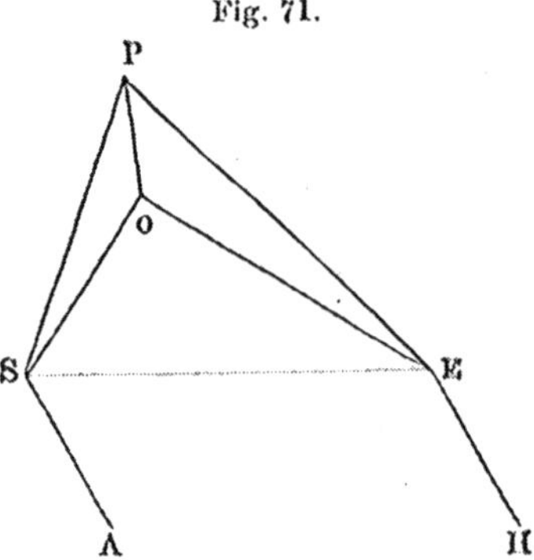

of the ecliptic (from H to O on the right side of the lines), be the geocentric longitude of the planet; then ASO will be its heliocentric longitude. Also PEO, the angular distance of the planet from the ecliptic in a plane perpendicular to it, is the geocentric latitude, and PSO is the heliocentric latitude. The planet's angular distance from the sun, PES, is also known from observation. Hence, in the triangle SEP, we know SP and SE and the angle SEP, from which we can find PE; and knowing PE and the angle PEO, in the right-angled triangle, we calculate EO. Next, in the triangle OES, EO and ES are known; also OES, found by subtracting the planet's longitude from the sun's, *i. e.*, HEO from HES (reckoned to the right from H); hence OSE and OS are calculated. If OSE be added to ASE, the supplement of HES, we have ASO, the *heliocentric longitude* of the planet. The line OS just found, is called the *curtate* distance of the planet from the sun; and this, with the *actual* distance SP, gives us, in the right-angled triangle SPO, the *heliocentric latitude* PSO. Thus, by a few processes in plane trigonometry, we can do what is equivalent to making a transfer of our position from the earth to the sun.

370. Having now learned how observations made at the earth may be converted into corresponding observations made at the sun, we may proceed to explain the mode of finding the several elements before enumerated; although our limits will not permit us to enter further into detail on this subject, than to explain the leading principles on which each of these elements is determined.*

371. First, to determine the *position of the Nodes*, and the *Inclination of the Orbit.*

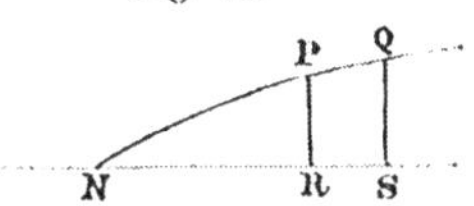

Fig. 72.

These two elements, which determine the situation of the orbit (Art. 367), may be derived from two heliocentric longitudes and latitudes. Let ANRS be an arc of the ecliptic, A the vernal equinox, NQ an arc of a planet's orbit, and N its node. Let AR, AS, be the two heliocentric longitudes; PR, QS, the corresponding heliocentric latitudes, which have been determined as by Art. 369.

By Napier's rule,

$$\sin NS = \tan QS \times \cot PNR,$$
$$\text{and } \sin NR = \tan PR \times \cot PNR.$$

Eliminating cot PNR, we have

$$\frac{\sin NS}{\tan QS} = \frac{\sin NR}{\tan PR};$$

substituting AS − AN for NS, and AR − AN for NR, then

$$\frac{\sin AS \times \cos AN - \cos AS \times \sin AN}{\tan QS} =$$
$$\frac{\sin AR \times \cos AN - \cos AR \times \sin AN}{\tan PR}.$$

$$\therefore (\cos AR \times \tan QS - \cos AS \times \tan PR) \sin AN =$$
$$(\sin AR \times \tan QS - \sin AS \times \tan PR) \cos AN.$$

$$\therefore \frac{\sin AN}{\cos AN} (= \tan AN) = \frac{\sin AR \times \tan QS - \sin AS \times \tan PR}{\cos AR \times \tan QS - \cos AS \times \tan PR}.$$

Thus AN, the longitude of the node is found; for all the quantities on the second member of the equation are known.

Again, since AN is found, we may deduce from the first

* Most of these elements admit of being determined in several different ways, an explanation of which may be found in the larger works on Astronomy.

equation above, the value of PNR, which is the *inclination of the orbit.*

372. Secondly, to find the *Periodic Time.*

This element is learned, by marking the interval that passes from the time when a planet is in one of the nodes until it returns to the same node. We may know when a planet is at the node, because then its latitude is nothing. If, from a series of observations on the right ascension and declination of a planet, we deduce the latitudes, and find that one of the observations gives the latitude 0, we infer that the planet was at that moment at the node. But if, as commonly happens, no observation gives exactly 0, then we take two latitudes that are nearest to 0, but on opposite sides of the ecliptic, one south and the other north, and as the sum of the arcs of latitude is to the whole interval, so is one of the arcs to the corresponding time in which it was described, which time being added to the first observation, or subtracted from the second, will give the precise moment when the planet was at the node.

By repeated observations it is found, that the nodes of the planets have a very slow retrograde motion.

373. If the orbit of a planet cut the ecliptic at right angles, then small changes of place would be attended by appreciable differences of latitude; but in fact the planetary orbits are in general but little inclined to the ecliptic, and some of them lie almost in the same plane with it. Hence arises a difficulty in ascertaining the exact time when a planet reaches its node. Among the most valuable observations for determining the elements of a planet's orbit, are those made when a superior planet is in or near *its opposition* to the sun, for then the heliocentric and geocentric longitudes are the same. When a number of oppositions are observed, the planet's motion in longitude, as would be observed from the sun, will be known. The inferior planets also, when in superior conjunction, have their geocentric and heliocentric longitudes the same. When in inferior conjunction, these longitudes differ 180°; but the inferior planets can seldom be observed in superior conjunction, on account of their proximity to the sun, nor in inferior conjunction except in their transits, which occur too rarely to admit of observations sufficiently numerous. Therefore, we can not so readily ascertain by simple observation, the motions of the inferior planets seen from the sun, as we can those of the superior.*

374. Hence, to obtain accurately the periodic time of a superior planet, we find the interval elapsed between two oppositions separated by a long interval, when the planet was nearly in the same part of the zodiac. From the periodic time, as determined approximately by other methods, it may be found when the planet has the same heliocentric longitude as at the first observation. Thus the time of a complete number of revolutions will be known, and thence the time of one revolution. The greater the interval of time between the two oppositions, the more accurately the periodic time will be obtained, because the errors of observation will be divided between a great number of periods; therefore by using very accurate observations, much precision may be attained. For example, the planet Saturn was observed in the year 228 B. C., March 2 (according to our reckoning of time), to be near a certain star called γ Virginis, and it was at the same time nearly in opposition to the sun. The same planet was again observed in opposition to the sun, and having nearly the same longitude, in Feb., 1714. The exact difference between these dates was 1943y. 118d. 21h. 15m. It is known from other sources, that the time of a revolution is 29½ years nearly, and hence it was found that in the above period there were 66 revolutions of Saturn; and dividing the interval by this number, we obtain 29.444 years, which is nearly the periodic time of Saturn according to the most accurate determination.

375. Thirdly, to determine *the distance from the sun, and major axes of the planetary orbits.*

The distance of the earth from the sun being known, the *mean* distance of any planet (its periodic time being known) may be found by Kepler's law, that the squares of the periodic times are as the cubes of the distances. The method of finding the distance of an *inferior* planet from the sun by observa-

---

* Brinkley, p. 167.

tions at the greatest elongation, has been already explained (see Art. 308). The distance of a *superior* planet may be found from observations on its retrograde motion at the time of opposition. The periodic times of two planets being known, we of course know their mean angular velocities, which are inversely as the times. Therefore, let E*e* (Fig. 73) be a very small portion of the earth's orbit, and M*m* a corresponding portion of that of a superior planet, described on the

Fig. 73

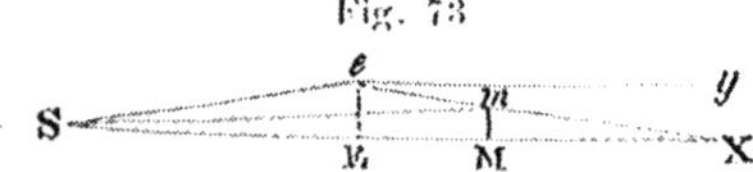

day of opposition, about the sun S, on which day the three bodies lie in one straight line SEMX. Then the angles ES*e* and MS*m*, representing the respective angular velocities of the two bodies, are known. Now if *em* be joined, and prolonged to meet SM continued in X, the angle EX*e*, which is equal to the alternate angle X*ey*, being equal to the retrogradation of the planet in the same time (being known from observation), is also given. E*e*, therefore, and the angle EX*e* being given in the right-angled triangle EX*e*, the side EX is easily calculated, and thus SX becomes known. Consequently, in the triangle S*m*X, we have given the side SX, and the two angles *m*SX and *m*XS, whence the other sides S*m* and *m*X are easily determined. Now S*m* is the radius vector of the orbit of the superior planet at the point through which it was passing at the time of the observation. Of course, *one* such observation could not be relied on as giving the *mean* distance; but it would be a satisfactory *approximation* in the case of any planetary orbit, since these orbits are all very nearly circular. And by repeating the process every year, as the earth passes between the sun and planet, the average of all will ultimately express the mean distance, or semi-major axis of the orbit in question.*

376. Fourthly, to determine the *place of the perihelion*, the *time of passing it*, and *the eccentricity*.

A method applicable to the inferior planets, is to make a

* Sir J. Herschel.

series of observations upon them at the times of their greatest elongations, as described in Art. 308. If the length of the radius vector be thus obtained at each return to the point of greatest elongation, there will be found among them all, one that is a maximum, and another that is a minimum. The latter point is approximately the place of the perihelion. Thus (Fig. 60), if in a long series of observations on the greatest elongations of Mercury, the value of SB were, at a certain time, to be the least of all, we should know that that point is the place of perihelion, and, of course, that the point diametrically opposite is the place of the aphelion. Moreover, by calculating the distances of the planet from the sun at these two points, as described in Art. 308, we ascertain the length of the least and greatest radius vector; and half the difference of these two lines constitutes the eccentricity.

For the superior planets, we might suppose the method described in Art. 375, to be pursued every year at the time of opposition, till a radius vector is found, which is less than any other; that is approximately the place of perihelion. For the most distant planets, however, the angle of retrogradation Xey, or eXS, is so minute, that the calculated value of Sm is very uncertain. Moreover, the distant planets pass their aphelion and perihelion at long intervals. So that, in a given case, it may be necessary to wait 40 years for Uranus, or 80 for Neptune, to pass either of the apsides.

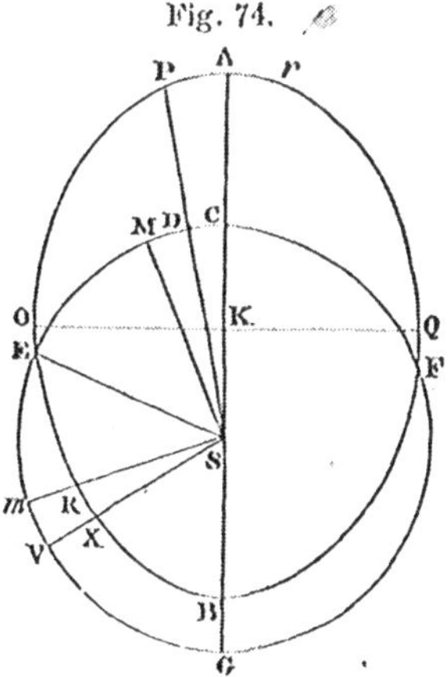

Fig. 74.

But trigonometry, building on a few instrumental observations, affords other modes of arriving at these elements of a planetary orbit, one of which is derived from the *greatest equation of the center* (Art. 200). For since the two points in the orbit where this becomes greatest are equally distant from the apsides, by bisecting the interval between these two points, we obtain the position of the perihelion and aphelion. Let AEBF (Fig. 74) be the orbit of the planet, having the sun in the focus at S. In an ellipse, the square root of the product of the semi-axes gives the radius of a circle of the same area as the ellipse. Therefore, with the center S, at the distance SE= $\sqrt{AK \times OK}$, describe the circle CEGF, then will the area of this circle be equal to that of the ellipse. At the same time that a body departs from A the aphelion, let a body begin to move with a uniform motion from C through the periphery CEGF, and perform a whole revolution in the same period that the planet describes the ellipse; the motion of this body will represent the equable or mean motion of the planet, and it will describe around S areas or sectors of circles which are proportional to the times, and equal to the elliptic areas described in the same time by the planet. Suppose the body describing the circle to be at M; then taking the sector ASP =CSM, P is the true place of the planet. The angle CSM is is the mean anomaly,* CSD the true, and DSM the equation of the center. But in a circle, *sectors* vary as their *arcs;* therefore, the sector DSM may be used for the equation; but taking CSD from the equals CSM and ASP, DSM=ACDP, which therefore measures the equation of the center. Now ACDP obviously *increases* till it becomes ACE, that is, when the planet has reached the point where the two orbits intersect. It may be shown, that after passing E, the equation *diminishes.* The half orbits AEB, CEG, are described in the same time; and the mean place, therefore, remains in advance of the true, till they reach G and B together. Let V be the mean place, and R the true place, at a certain moment; then the angle CSV is the mean anomaly, and CS*m* the true, and VS*m* the equation. The sectors ASR, CSV, are equal, being described in the same time. Taking CERS from each, ACE= E*m*R+VS*m*; ∴ VS*m*=ACE−E*m*R; that is, the equation VS*m* has diminished by E*m*R since the planet was at E. Therefore E is the place of greatest equation of the center.

But E is also the place where the mean and true angular motions are equal, because the equal sectors of the two orbits described in each instant, have the same length at that point, and therefore the same angle. Hence, the greatest equation occurs *where the mean angular motion is equal to the true.*

---

* Anomaly is now reckoned from *perihelion* (Art. 200); but that change does not affect the correctness of this reasoning.

The mean angular motion being known from the periodic time of the planet, it is then ascertained by observation when the true motion equals it, and thus the time of the greatest equation of the center is obtained. Now, this occurs twice in the revolution, at E and F; and half way between these points lie the apsides A and B. Therefore, observing the *times* of the greatest equation of the center, E and F, and bisecting the interval, we have the *time of the planet's* passing *the perihelion* B. But the same observations also determine the heliocentric *places* of E and F, and the middle of the arc EBF is the *place of the perihelion.*

377. The amount of the greatest equation evidently depends on the *eccentricity* of the orbit, since it arises wholly from the departure of the ellipse from the figure of a perfect circle; hence, the greatest equation affords the means of determining the eccentricity itself. In orbits of small eccentricity, as is the case with most of the planetary orbits, it is found that the arc which measures the greatest equation is very nearly equal to the distance between the foci, which always equals twice the eccentricity, the measure of the eccentricity being the distance from the focus to the center of the ellipse. The angular value of radius is 57° 17′ 44″.8; for,

3.14159 : 1 :: 180° : 57° 17′ 44″.8.

Therefore, 57° 17′ 44″.8 : radius :: half the greatest equation of the center : the *eccentricity.**

The foregoing explanations of the methods of finding the elements of the orbits, will serve in general to show the learner how these particulars are or may be ascertained: yet the methods actually employed are usually more refined and intricate than these. In astronomy, scarcely an element is presented simple and unmixed with others. Its value, when first disengaged, must partake of the uncertainty to which the other elements are subject, and can be supposed to be settled to a tolerable degree of correctness, only after multiplied observations and many revisions.† Indeed, a large part of the most arduous labors of astronomers have been employed in finding

* Vince's Complete System, i., p. 113. † Woodhouse, p. 579.

the elements of the planetary orbits, with the wonderful degree of precision which has finally been attained.

QUANTITY OF MATTER IN THE SUN AND PLANETS.

378. It would seem at first view very improbable, that an inhabitant of this earth would be able to weigh the sun and planets, and estimate the exact quantity of matter which they severally contain. But the principles of Universal Gravitation conduct us to this result, by a process remarkable for its simplicity. By comparing the relations of a few elements that are known to us, we ascend to the knowledge of such as appeared beyond the pale of human investigation. *We learn the quantity of matter in a body by the force of gravity it exerts.* Let us see how this force is ascertained.

379. *The quantities of matter in two bodies may be found in terms of the distances and periodic times of two bodies revolving around them respectively, being as the cubes of the distances divided by the squares of the periodic times.*

The force of gravity G in a body whose quantity of matter is M and distance D, varies directly as the quantity of matter, and inversely as the square of the distance; that is, $G \propto \frac{M}{D^2}$. But it is shown of circular orbits (Art. 177), that the force of gravity also varies as the distance divided by the square of the periodic time, or $G \propto \frac{D}{P^2}$. Therefore, $\frac{M}{D^2} \propto \frac{D}{P^2}$, and $M \propto \frac{D^3}{P^2}$. Thus we may find the respective quantities of matter in the earth and the sun by comparing the distance and periodic time of the moon revolving around the earth, with the distance and periodic time of the earth revolving around the sun. For the cube of the moon's distance from the earth divided by the square of her periodic time, is to the cube of the earth's distance from the sun divided by the square of her periodic time, as the quantity of matter in the earth is to that in the sun. That is, $\frac{238{,}545^3}{27.32^2} : \frac{95{,}000{,}000^3}{365.256^2} :: 1 : 353{,}385$. The most exact determination of this ratio gives for the mass of the sun

354,936 times that of the earth. Hence it appears that the sun contains more than three hundred and fifty-four thousand times as much matter as the earth. Indeed, the sun contains eight hundred times as much matter as all the planets.

Another method, well suited to popular illustration, of weighing the earth against the sun, is the following. Knowing the radii of the solar and lunar orbits respectively, we can easily find the space which the moon descends toward the earth, and the earth toward the sun, in any given time, as an hour. Thus (Fig. 75), if we know the radius AE of the orbit, we can

Fig. 75.

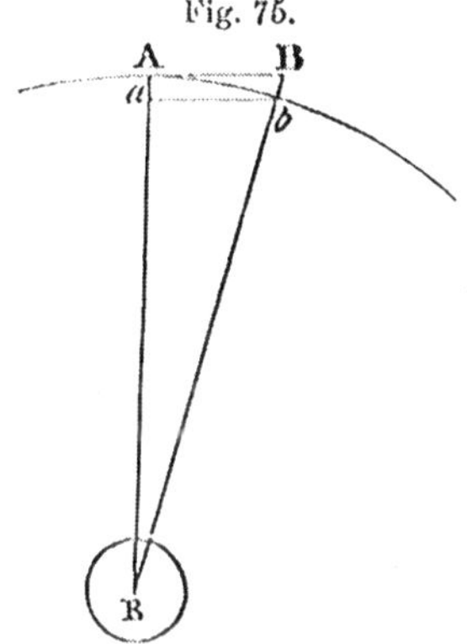

determine the length of the arc Ab, described in an hour, and also the length of the hypotenuse BE. But BE—AE=Bb, the space through which the central attracts the revolving body in the given time. It was shown (Art. 182) that the earth draws the moon from a tangent, .0536 of an inch in a second: if a calculation of the same kind be made in relation to the orbit of the earth, it will be found that the sun draws the earth nearly .12 of an inch per second from a tangent; that is, the sun exerts a force $2\frac{1}{5}$ greater on the earth than the earth does on the moon. But were the sun at the same distance as the moon, his force of attraction would be the square of 400, or 160,000 times as great as it is now; that is, it would be $2\frac{1}{5}\times160{,}000$ times as great as the earth's attraction, and, consequently, must have $2\frac{1}{5}\times160{,}000=352{,}000$ times as much matter—a result agreeing nearly with the former. The agreement would be exact if more precise numbers were employed, but our object is here merely to illustrate the method.

380. The mass of each of the other planets *that have satellites* may be found by comparing the periodic time of one of its satellites with its own periodic time around the sun. By this means we learn the ratio of its quantity of matter to that of the sun. The masses of those planets which have no satellites, as Venus or Mars, have been determined by estimating the force of attraction which they exert in disturbing the motions of other bodies. Thus, the effect of the moon in raising the tides, leads to a knowledge of the quantity of matter in the moon; and the effect of Venus in disturbing the motions of the earth, indicates her quantity of matter.*

381. The quantity of matter in bodies varies as their magnitudes and densities conjointly. Hence, their *densities* vary as their masses divided by their magnitudes; and since we know the magnitudes of the planets, and can compute as above their masses, we can thus learn their densities, which, when reduced to a common standard, give us their specific gravities, or show us how much heavier they are than water. Worlds, therefore, are weighed with the same certainty as a pebble, or an article of merchandise.

The *densities and specific gravities* of the sun, moon, and planets, are estimated as follows:†

	Density.	Specific Gravity.
Sun,	0.25	1.37‡
Moon,	0.56	3.07
Mercury,	1.12	6.13
Venus,	0.92	5.04
Earth,	1.00	5.48
Mars,	0.95	5.20
Jupiter,	0.24	1.31
Saturn,	0.14	0.76
Uranus,	0.24	1.31
Neptune,	0.14	0.76

* These estimates are made by the most profound investigations in Laplace's Mécanique Céleste, vol. iii.

† Herschel.

‡ The earth being taken, according to Baily, at 5.48, the specific gravities of the other bodies (which are found by multiplying the density of each by the specific gravity of the earth) are here stated somewhat higher than they are given in most works.

From this table it appears that the sun consists of matter but little heavier than water; but that the moon is more than three times as heavy as water, though less dense than the earth, which is five and a half times heavier than water. It also appears that the nearer planets are more dense than the more remote. Mercury is heavier than most metallic ores, while Saturn and Neptune are one-fourth lighter than water. The density, however, does not, in all cases, diminish outward; for Venus is less dense than the earth, and Saturn than Uranus.

CHAPTER XII.

PERTURBATIONS OF THE PLANETS—STABILITY OF THE SYSTEM—NUMERICAL RELATIONS OF THE PLANETS—PROBLEMS.

382. The *perturbations* occasioned in the motions of the planets by their action on each other are very numerous, since every body in the system exerts an attraction on every other, in conformity with the law of universal gravitation. Venus and Mars, approaching as they do at times comparatively near to the earth, sensibly disturb its motions; and Jupiter and Saturn, although very far asunder, still, in consequence of their great masses, exert on each other, when on the same side of the heavens especially, a decided influence. Moreover, the sun, by his *unequal* action on the several planets, in consequence of the peculiar figure of each, produces various irregularities in their motions. As in the case of the earth and moon (Art. 243), these perturbations are divided into periodical and secular: *periodical*, when completed in comparatively short periods, as those, for example, which undergo all their changes during one revolution of the planet; and *secular*, when completed only in very long periods, as those which affect the form and inclination of the orbits.

383. If the only bodies in the system were a central body like the sun, and a revolving body like Venus, then, when the planet was once put in motion with such a projectile force as to make it describe an ellipse, it would forever continue to describe the same figure without the least variation, the radius vector always passing over equal spaces in equal times; but now introduce a third body so near as to exert on it a decided attraction, and its motions no longer retain their simplicity, but become complicated by the conflicting influences of the two attracting bodies. The sun, however, in consequence of its mass, which is eight hundred times as great as that of all the planets, and, of course, vastly greater than that of any one of them, exerts a force so much superior to that of any or all the other disturbing bodies, that the elliptical figure of the orbits is *nearly* maintained, and a near approximation to the place of a planet is obtained by neglecting all those minor forces, and simply contemplating it as revolving in an elliptical orbit. Still it is essential, in order to find the *exact* place of a planet at any given time, that all these irregularities, minute as they may be, be carefully summed up, and their resultant applied to the elliptical motions. To investigate these perturbations, to estimate their precise amount, and to register them in tables, for the use of the practical astronomer, have constituted a large part of the labors of modern astronomy. The knowledge gained by astronomers of the planetary motions, considering the very numerous irregularities, both periodical and secular, to which they are subject, is truly wonderful. The motion of Jupiter, for instance, is so perfectly calculated, that astronomers have computed ten years beforehand the time at which it will pass the meridian of different places, and we find the prediction correct within half a second of time.* The more obvious irregularities have been detected by observation; the more minute, by following out the consequences of universal gravitation. Even those at first revealed to the instruments of the astronomer have been confirmed and estimated with greater accuracy by the same far-reaching principle; and many of the irregularities have been first brought to light by this theory, which had before eluded observation; although, when once pointed out as a result of the principle of gravitation, careful instrumental measurements

* Airy.

have confirmed them, except in cases where the force was too minute to be reached by the most refined observation. Periodical perturbations among the bodies of the solar system may be compared to the regular flux and reflux of the tides, by which the ocean daily oscillates about its mean level, without any permanent change of level; while secular perturbations would resemble any slow changes of level, which, accumulating from time to time, might finally become obvious to measures of the depths of the ocean, as recorded from age to age. As an example of the extreme minuteness of some of these secular perturbations, we may instance the changes in the eccentricity of the earth's orbit. The entire eccentricity is so small that the figure, when drawn on paper in just proportions, can scarcely be distinguished from a circle, the focus of the ellipse being distant from the center only about $\frac{1}{60}$ part of the semi-major axis. But the *change* of eccentricity in a *century* is only the twenty-five thousandth part of the whole, or the fifteen hundred thousandth part of the semi-major axis.

384. But although the secular inequalities of the planetary motions are exceedingly slow, yet may they not, in time, accumulate so as to derange the whole system; and do they not, at least, indicate that the system carries within it the seeds of its own dissolution? So far is this from being the case, that the *stability of the solar system* is a fact established on the most satisfactory evidence, and its demonstration is among the finest triumphs of physical astronomy. Even a superficial view of the system will convince us that care has been bestowed on this point by several obvious arrangements. One is, that the planets have, severally, so small masses, compared with the sun, as to interfere but little, at most, with the supremacy of his control over the planetary motions. Another is, that the planets are placed at such great distances from each other—a distance which is greater among the largest bodies, as Jupiter and Saturn, than among the smaller, as the Earth and Venus; and another still, that the orbits are less eccentric when the masses of the bodies are greater, by which provision they are always maintained at a remote distance from the sun. Were the orbit of Jupiter as eccentric as that of Mars, he would approach so near the earth at his perihelion, as greatly to endanger its stability. But if even these general considerations might convince us that the stability of the solar system is provided for, a more profound investigation will reveal this truth in a far more admirable light. This object is especially secured by the following remarkable provisions.

First, by the *invariability of the major axes*, and of the *periodic times;* secondly, by the fact, that whatever irregularities a planet undergoes on one side of its orbit (so far as respects the *periodical* perturbations), they are compensated on the other side; so that, when it returns to a given point, as the node or the perihelion, any irregularities it may have felt in different parts of its orbit neutralize one another, and therefore do not constitute an accumulating mass of errors; and, thirdly, by this, that all the secular perturbations are restricted within narrow limits, oscillating to and fro; but, before they can proceed so far on one side as to endanger the stability of the system, they turn about and proceed, for a similar period, in the opposite direction.

385. These truths have been established by the most rigorous mathematical demonstrations, by the successive labors of three very celebrated mathematicians—Euler, Lagrange, and Laplace. It was demonstrated that the major axes of the planetary orbits, and the times of their revolutions around the sun, are subject to no secular perturbations, nor to any variation whatever, but such as, in the course of a single revolution, exactly compensate and neutralize each other. This is a most important point in relation to the stability of the system; for if the lengths of the major axes varied, then, of course, the times of revolution would vary (since, by Kepler's third law, the squares of the periodic times are in a constant ratio to the cubes of the major axes), and we should have years of unequal length, and the earth, by approaching at one time nearer to the sun, and at another receding further from it, would render the changes of temperature too great for the existence of animal or vegetable life; and similar evils, it is probable, would result to the economy of the other planets. It was next established, that the *eccentricities* of the planetary orbits, although they have been undergoing constant changes in all time past, and will continue to undergo them in all future ages, can never vary beyond a certain moderate limit, entirely within the bounds of safety to the stability of the system. The eccentricity of the earth's orbit, for example, has been diminishing ever since the creation of man; and although, as we have seen, the rate of diminution is exceedingly slow, yet, in the progress of centuries, it would totally change the character of the earth's orbit; first reducing it to the circular form, and finally carrying its eccentricity to a fatal extreme. In like manner, the inclination of the earth's orbit to the equator is constantly diminishing, and is now about two-fifths of a degree less than it was in the days of Aristotle; and, were this to proceed in the same direction, the equator and ecliptic would coincide, the change of seasons would cease, and the whole economy of nature would be subverted. But Laplace has demonstrated, that such an event can never occur, nor can the entire extent of this variation exceed three degrees. It is worthy of remark, that those perturbations, such as changes in the place of the perihelion, affecting a change of direction in space of the major axis of the orbit, or in the place of the nodes, which, by accumulating, do not endanger the stability of the system, proceed onward through the entire circuit of the heavens; while perturbations which, by indefinite accumulation, would bring ruin to the system, such as variations of eccentricity and of inclination, are not progressive, but oscillatory, waving to and fro within the limits of entire safety.

386. These great ends would not have been secured, had the system been constructed differently from what it is. Numerous conditions must concur in order to produce these results: the mass of the sun must have greatly exceeded that of any or all the planets; the eccentricities of the orbits must have been small; and the planets must all have revolved around the sun in the same direction, and in planes but little inclined to each other.* It was also necessary that the periodic times of the planets should, in general, be incommensurable; for were their periods such that one planet would revolve a certain

* Laplace, Sys. du Monde. Herschel's outlines. Grant's History of Physical Astronomy. Pontecoulant's Trait. Elémen. de Phys. Céleste.

number of times exactly, while another planet, next to it, revolved a certain other even number of times, then, when they once came into the sphere of each other's influence, they might remain under it so long, and return to their relative position so often, as seriously to derange their orbits. An instance of this, in fact, occurs in the case of Jupiter and Saturn, five revolutions of Jupiter being nearly equal to two of Saturn, a relation which gives rise to what is called the *long inequality of Saturn and Jupiter*. Similar effects result from a near commensurability of the mean motions of any other two planets. One exists between the Earth and Venus, 13 times the period of Venus being very nearly equal to 8 times that of the Earth; still, the influence of this disturbing cause is so nicely compensated, and its effects so distributed, that, according to Mr. Airy (who was the first to detect it), it amounts, at its maximum, to no more than a few seconds for a period of 240 years. The laws which regulate the eccentricities and inclinations of the planetary orbits (says an able writer on Physical Astronomy), combined with the invariability of the mean distances, secure the permanence of the solar system throughout an indefinite lapse of ages, and offer to us an impressive indication of the Supreme Intelligence which presides over nature, and perpetuates her beneficent arrangements. When contemplated merely as speculative truths, they are unquestionably the most important which the transcendental analysis has disclosed to the researches of the geometer; and their complete establishment would suffice to immortalize the names of Lagrange and Laplace, even although these great geniuses possessed no other claim to the recollection of posterity.*

NUMERICAL RELATIONS BETWEEN THE BODIES OF THE SOLAR SYSTEM.†

387. If we contemplate the relations subsisting between a central body, as the sun, and a revolving body, as one of the

* Grant's Hist. Phys. Ast., p. 56.

† In the preparation of this article, the author has derived much assistance from a small work, now nearly out of print, containing the substance of three lectures delivered to the students of Yale College in 1781, by Rev. Nehemiah Strong, at that time Professor of Mathematics and Natural Philosophy.

planets, it will be readily understood, that if the quantity of matter in the central body is increased, while the distance of the revolving body remains the same, the velocity of the revolving body must be increased also, in order to generate a sufficient centrifugal force to counterbalance the increased force of attraction in the central body, arising from the increase of its mass; and that, were the force of attraction diminished by removing the body to a greater distance from the center, then the rate of its motion would also have to be diminished, otherwise the centrifugal force would overpower the force of attraction. It is a remarkable fact, that the members of the solar system are so adjusted to each other, in respect to their velocities, distances from the sun, periodic times, and gravitation toward the central body, that if any one of these particulars is known, all the rest become known also. Thus, if it were found that a new-discovered planet moved with one-sixth the velocity of the earth, we should know at once that its distance from the sun was thirty-six times as great as the earth's distance, that its time of revolution was two hundred and sixteen years, and that its gravitation toward the sun was twelve hundred and ninety-six times less than that of the earth; for the distance is as the *square* of the number expressing the reciprocal of its velocity, compared with that of the earth; its periodic time as the *cube;* and the reciprocal of gravity as the *fourth power* of the same number. All this follows from Kepler's third law—that the squares of the periodic times are as the cubes of the distances; and from the law of universal gravitation—that the force of attraction is inversely as the square of the distance. The four particulars named, therefore, constitute a series of numbers in geometrical progression, of which the first term is equal to the ratio. The truth of this proposition may be demonstrated as follows:

Let D be the mean distance of a planet from the sun, π the ratio of the diameter to the circumference of a circle, and P the time of revolution around the sun, or periodic time; then the expression for the velocity is $V = \frac{2\pi D}{P} \propto \frac{D}{P}$. And $V^2 \propto \frac{D^2}{P^2}$. But, by Kepler's law, $P^2 \propto D^3$; $\therefore V^2 \propto \frac{D^2}{D^3}$, or $V^2 \propto \frac{1}{D}$. Since a body moves with *less* velocity, when the distance from the sun is *greater*, it will be convenient, in order to avoid fractional forms, to use the reciprocal of V, instead of V itself. Let R (*retardation*) be equal to $\frac{1}{V}$; then $V = \frac{1}{R}$, and $V^2 = \frac{1}{R^2}$; hence $\frac{1}{R^2} \propto \frac{1}{D}$; $\therefore R^2 \propto D$ (1). If, therefore, R indicates how much slower a planet moves than another, as the earth, taken as a standard, the square of R will show how much further from the sun the planet is than the earth.

Again, since $V \propto \frac{D}{P}$, $V^3 \propto \frac{D^3}{P^3}$. But, by Kepler's law, $D^3 \propto P^2$; $\therefore V^3 \propto \frac{P^2}{P^3}$, or $V^3 \propto \frac{1}{P}$; $\therefore R^3 \propto P$ (2).

Consequently, if R expresses how many times slower than the earth a given planet moves, the cube of R will express the relative periodic time.

Finally, by the law of gravitation, the force of gravitation toward the central body varies as the square of the distance inversely, or $G \propto \frac{1}{D^2}$. But if the reciprocal of gravity be called LEVITY, and expressed by L, then $L \propto D^2$; but $R^2 \propto D$, $\therefore R^4 \propto D^2$, and $R^4 \propto L$ (3).

Therefore, if R denotes how much slower a planet moves in its orbit than the earth, R^4 will denote how much less the same body gravitates toward the central body. Collecting these several results, it appears that *the reciprocal of velocity* R, *the distance* D, *the periodic time* P, and *the reciprocal of gravity* L, are respectively denoted by the geometrical series, R, R^2, R^3, R^4, in which the first term and the ratio are equal.

388. A number of very useful and convenient rules, may be derived from this numerical relation between the members of the solar system; since, when any one of the four things named is given, all the rest may be found from it; and each of the four may be found in four different ways when the other members of the series are given. This will be obvious from a few examples.

I. Given the RETARDATION (R).

1. Square the retardation for the *distance.*

2. Cube the retardation for the *periodic time.*

3. Take the fourth power of the retardation for the *reciprocal of gravity.*

II. Given the DISTANCE (D).

1. Take its square root for *reciprocal of velocity.*

2. Take the cube of the square root of the distance for the *periodic time.*

3. Take its square for the *reciprocal of gravity.*

III. Given the PERIODIC TIME (P).

1. Take the cube root of the periodic time for the *reciprocal of velocity.*

2. Take the square of the cube root of the periodic time for the *distance.*

3. Take the fourth power of its cube root for the *reciprocal of gravity.*

IV. Given the RECIPROCAL OF GRAVITY (L).

1. Take the fourth root for the *reciprocal of velocity.*

2. Take the square root for the *distance.*

3. Take the cube of the fourth root for the *periodic time.*

V. Required the RECIPROCAL OF VELOCITY.

This may be obtained by taking the square root of the distance, or the cube root of the periodic time, or the fourth root of the reciprocal of gravity, or by dividing the reciprocal of gravity by the periodic time.

VI. Required the DISTANCE.

Take the square of the reciprocal of velocity, or the square of the cube root of the time, or the square root of the reciprocal of gravity, or divide the time by the reciprocal of velocity.

VII. Required the PERIODIC TIME.

Take the cube of the reciprocal of velocity, or the cube of the square root of the distance, or the $\frac{3}{4}$ power of the reciprocal of gravity, or divide the reciprocal of gravity by that of velocity.

VIII. Required the diminished GRAVITATION.

Take the fourth power of the reciprocal of velocity, or the square of the distance, or the $\frac{4}{3}$ power of the time, or multiply the time by the reciprocal of velocity.

According to the foregoing rules, *tables* may be formed, exhibiting, in a striking light, the numerical relations of the members of the solar system. In the following table the *distances* are taken from Herschel's Astronomy, and from these the other particulars are determined by the preceding rules. If Mercury were taken as the standard of comparison, then the retardations of all the other planets would be greater than unity; but, as it is convenient to take the earth as the standard, the retardations of Mercury and Venus will be less than unity; showing that the velocity (which is expressed by the fraction inverted) is greater than that of the earth. In like manner, the force of gravitation of an inferior planet, being greater than that of the earth, is the reciprocal of the tabular number.

TABLE SHOWING THE NUMERICAL RELATIONS OF THE PRIMARY PLANETS.

Planets.	Retardations.	Distances.	Per. Times.	Recip. of Gravity.
Mercury	0.62217	0.38710	0.24084	0.14985
Venus	0.85049	0.72333	0.61519	0.52321
Earth	1.00000	1.00000	1.00000	1.00000
Mars	1.23440	1.52369	1.88080	2.32170
Jupiter	2.28100	5.20277	11.86700	27.06900
Saturn.......	3.08850	9.53878	29.46100	90.98900
Uranus	4.37970	19.18239	84.01200	367.95000
Neptune	5.49040	30.14512	165.51000	908.72000

389. PROBLEMS.

PROB. 1.—The planet Pallas was discovered to have a period of about $4\frac{2}{3}$ years. How much slower does it move in its orbit than the earth—how much further is it from the sun—and how much less does it gravitate toward the sun? *Ans.* R = 1.67, D = 2.79, L = 7.80.

By applying the proportional numbers determined by this problem respectively to the earth's motion per second, to its distance from the sun in miles, and to the space through which the earth departs in a second from a tangent to her orbit, we may obtain the numerical value of each of these elements.

PROB. 2.—What would be the *periodical time* of a meteor or planet revolving close to the earth.

As the moon is a body revolving around the earth at a known distance, and with a known periodic time, it will evidently furnish the necessary standard of comparison. The distance of the moon from the center of the earth being 60 times the earth's radius, and, of course, 60 times that of the meteor, its rate of motion is $\sqrt{60}$ times less. The retardation being $\sqrt{60}$, the periodic time will be $60^{\frac{3}{2}}$. Now, what part of the moon's period is $60^{\frac{3}{2}}$? Divide the moon's period (27.32 days) by $60^{\frac{3}{2}}$, and we have for the answer, 1 hour, 24 minutes, 38.88 seconds.

PROB. 3.—What would be the periodic time of a body revolving about the earth at the distance of 5000 miles from the center? *Ans.* 1h. 59m. 23.28s.

PROB. 4.—How much faster must the earth revolve in order that bodies on its surface may lose all their gravity?

According to problem 2, the period of a body revolving at the surface of the earth, is 1.4108 hours; and since, in a circular orbit, the force of gravity and the centrifugal force are equal, therefore a body like that contemplated in problem 2, is in equilibrium between these two forces; consequently, such a body may be considered as having lost all its gravity, and being, by the supposition, close to the earth, we have only to inquire how much its velocity exceeds that of the earth. Now, 24 divided by 1.4108 gives 17.01; which shows that were the earth to revolve on its axis about 17 times faster than it does at present, the bodies on the surface would lose all their weight; and were the velocity greater than this, the centrifugal force would prevail over the centripetal, and the same would fly off from the earth in tangents.

PROB. 5.—Were the moon to be removed so far from the earth as to revolve about it but once a year, how much greater would be its distance than at present, how much less its velocity, and its gravitation toward the earth?

Its period being increased 13.37 times, its retardation is $13.37^{\frac{1}{3}}=2.373$; its distance $2.373^2=5.631$; and its diminished gravity $5.631^2=31.71$. Or $R=2.373$, $D=5.631$, and $L=31.71$.

Multiplying the present distance of the moon, 238,545 miles, by 5.631, we obtain about 1,343,000 miles for the distance at which the moon must have been placed in order to complete its revolution in one year.

PROB. 6.—Were the earth's mass equal to the sun's, and of course 354,000 times as great as at present, in what time would the moon revolve around it?

Since the masses are as the cubes of the distances divided by the squares of the periodic times, letting the required time be denoted by x, 1 (the earth's mass) : 354.000 (the sun's mass) $:: \frac{D^3}{27.32^2} : \frac{D^3}{x^2} :: \frac{1}{27.32^2} : \frac{1}{x^2}, \therefore \frac{1}{x^2} = \frac{354,000}{27.32^2}, \therefore x = \frac{27.32}{\sqrt{354,000}} =$ 1h. 6m. 7s.

Comets, in passing their perihelion, especially when that happens to be very near the sun, as in the great comet of 1843, move with an astonishing rapidity; requiring a velocity not merely sufficient to generate the centrifugal force necessary to balance the powerful force of attraction exerted by the sun, but greatly to exceed that force, since they are carried far without a circular orbit into an elliptical or even a hyperbolic orbit.

PROB. 7.—The perihelion distance of the great comet of 1843 being 502,000 miles from the center of the sun, what must have been its velocity per hour, if in a circular orbit?

PROB. 8.—How much must the mass of the earth be increased in order that the moon may revolve about it in the same time as at present, when removed to three times her present distance?

PROB. 9.—How much must the mass of the earth be increased to make the moon, at her present distance, revolve in 24 hours?

PROB. 10.—The semi-diameter of Jupiter being 11 times that of the earth, and the distance of its fourth satellite from the center of the planet being 27 times the radius of the planet; also the sidereal revolution of the satellite being 16.69 days, while that of the moon is 27.3217 days, and her distance 60 times the radius of the earth: How much does the quantity of matter in Jupiter exceed that of the earth? *Ans.* 324.49 times.

PROB. 11.—Suppose volcanic matter to be thrown from the moon toward the earth, required the point where it would be in equilibrium between the two, the mass of the moon being one-eightieth that of the earth? *Ans.* 24,000 miles from the center of the moon, *nearly*.

PROB. 12.—Suppose that the only two bodies in the universe were a sphere two inches in diameter, of the same density with the earth, for the primary, and a material point for the satellite. What would be the periodic time of the satellite, at the distance of one foot, in a circular orbit? *Ans.* 2 days, 10 hours, 13 minutes.*

---

* The elements used in the solution of this problem are, for the diameter of the earth, 7912.4; for the distance of the moon 238,545 miles; and for its periodic time, 27.32 days. The solution, conducted in the ordinary mode, will be found susceptible of great abridgment. But the following ingenious method is still shorter. It was suggested to the author by one of his pupils, Mr. Samuel Emerson, of the class of 1818.

LEMMA. *The periodic times of two satellites revolving about primaries of equal densities, at distances which are equimultiples of their radii, are equal.*

Demonstration. Let

M, m = the masses of the two bodies respectively.

P, p = the periodic times.

R, r = the radii of the spheres.

D, d = the distances of their satellites.

Then, $M : m :: \frac{D^3}{P^2} : \frac{d^3}{p^2}$.

But since D and d are equimultiples of R, r, by some number n, therefore $D^3 = R^3n^3$, and $d^3 = r^3n^3$;

Hence, $M : m :: \frac{R^3n^3}{P^2} : \frac{r^3n^3}{p^2} :: \frac{R^3}{P^2} : \frac{r^3}{p^2}$. But, R^3 and $r^3 \propto M$ and m.

Therefore, $M : m :: \frac{M}{P^2} : \frac{m}{p^2}$, $\therefore \frac{M \times m}{p^2} = \frac{M \times m}{P^2}$, $\therefore P = p$.

The moon being distant 60.296 radii of the earth (as would result from the above elements), at the distance of 60.296 inches that of the small satellite from its primary would be the same multiple of its radius, and consequently, its periodic time the same. What then is its period at 12 inches?

$$27.32^2 : p^2 :: 60.296^3 : 12^3, \quad \therefore p = 2d.\ 10h.\ 13m.$$

Corollary.—If any two spheres of the same density be taken, the periodic times of satellites revolving about them *close to the surface*, will be the same in both; for the case becomes this when $n = 1$. Thus, the material point supposed in the above problem, will revolve about its little globe in the same time that the moon would revolve about the earth, both being situated close to the surfaces of their respective primaries.

CHAPTER XIII.

COMETS—METEORIC SHOWERS.

390. A COMET,* when perfectly formed, consists of three parts—the Nucleus, the Envelope, and the Tail. The *Nucleus*, or body of the comet, is generally distinguished by its forming a bright point in the center of the head, conveying the idea of a solid, or at least of a very dense portion of matter. Though it is usually exceedingly small when compared with the other parts of the comet, yet it sometimes subtends an angle capable of being measured by the telescope. The *Envelope* (sometimes called the *coma*) is a dense nebulous covering, which frequently renders the edge of the nucleus so indistinct, that it is extremely difficult to ascertain its diameter with any degree of precision. Many comets have no nucleus, but present only a nebulous mass extremely attenuated on the confines, but gradually increasing in density toward the center. Indeed, there is a regular gradation of comets, from such as are composed merely of a gaseous or vapory medium, to those which have a well-defined nucleus. In some instances on record, astronomers have detected with their telescopes small stars through the densest part of a comet. The *Tail* is regarded as an expansion or prolongation of the coma; and presenting, as it sometimes does, a train of appalling magnitude, and of a pale, portentous light, it confers on this class of bodies their peculiar celebrity.

391. The number of comets belonging to the solar system, is probably very great. Many, no doubt, escape observation by being above the horizon in the daytime. Seneca mentions, that during a total eclipse of the sun, which happened 60 years before the Christian era, a large and splendid comet suddenly made its appearance, being very near the sun. The elements of at least 180 comets have been computed, and arranged in a

---

* Κόμη, *coma*, from the *bearded* appearance of comets.

catalogue for future comparison.* Of these, *six* are particularly remarkable, viz., the comets of 1680, 1770, and 1843; and those which bear the names of Halley, Encke, and Biela. The comet of 1680 was distinguished not only for its astonishing size and splendor, but is remarkable for having been the first comet whose elements were determined on the sure basis of mathematics, as was done by Sir Isaac Newton, it having appeared in his time. The comet of 1770 is memorable for

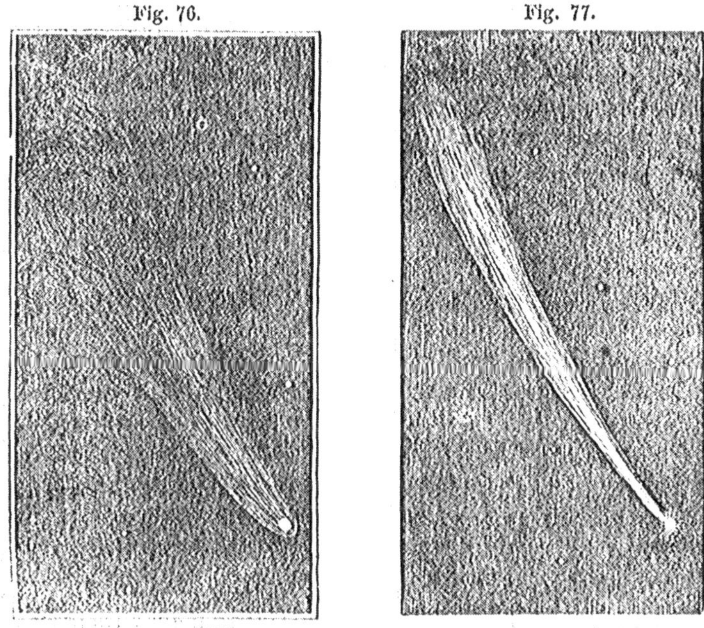

Fig. 76. COMET OF 1811.

Fig. 77. COMET OF 1680.

the changes its orbit has undergone by the action of Jupiter, and for having approached very near to the earth. The comet of 1843 was the most remarkable in its appearance of all that have been seen in modern times, having been visible at noonday. Halley's comet (the same which reappeared in 1835) is distinguished as that whose return was first successfully predicted, and whose orbit was first accurately determined; and Biela's and Encke's comets are well known for their short

* See a complete catalogue of comets, whose elements have been determined, in the American Almanac for 1847.

periods of revolution, which subject them frequently to the view of astronomers. Biela's comet, at its return in 1846, displayed another remarkable feature—*a separation into two distinct parts.* This strange peculiarity was first seen from the Observatory of Yale College, by Messrs. Herrick and Bradley, but was first publicly announced from the Observatory at Washington. At one time, the distance of one nucleus from the other, was estimated at 157,000 miles.

392. In *magnitude* and *brightness*, comets exhibit a great diversity. They are sometimes so bright as to be distinctly visible in the daytime, even at noon and in the brightest sunshine, as was the case with that of 1843; and such was the comet seen at Rome a little before the assassination of Julius Cæsar. The comet of 1680 covered an arc of the heavens of 97°, and its length was estimated at 123,000,000 miles.* That of 1811 had a nucleus of only 428 miles in diameter, but a tail 132,000,000 miles long.† Had it been coiled round the earth like a serpent, it would have reached round more than 5000 times. Other comets are of exceedingly small dimensions, the nucleus being estimated at only 25 miles; and some which are destitute of any perceptible nucleus, appear to the largest telescopes, even when nearest to us, only as a small speck of fog, or as a tuft of down. The majority of these bodies can be seen only by the aid of the telescope.

The same comet, indeed, has often very different aspects, at its different returns. Halley's comet in 1305 was described by the historians of that age, as *cometa horrendæ magnitudinis;* in 1456 its tail reached from the horizon to the zenith, and inspired such terror, that, by a decree of the Pope of Rome, public prayers were offered up at noonday in all the Catholic churches to deprecate the wrath of heaven, while in 1682, its tail was only 30° in length, and in 1759 it was visible only to the telescope, until after it had passed its perihelion. At its recent return in 1835, the greatest length of the tail was about 12°.‡ These changes in the appearances of the same comet are partly owing to the different positions of the earth with

* Arago. † Milne's Prize Essay on Comets.

‡ But might be seen much longer by *indirect* vision. (Prof. Joslin, Am. Journ. Science, xxxi., p. 328.)

respect to them, being sometimes much nearer to them when they cross its track than at others; also one spectator so situated as to see the comet at a higher angle of elevation or in a purer sky than another, will see the train longer than it appears to one less favorably situated; but the extent of the changes are such as indicate also a real change in their magnitude and brightness.

393. The *periods* of comets in their revolutions around the sun, are equally various. Encke's comet, which has the shortest known period, completes its revolution in 3⅓ years, or more accurately, in 1205.23 days; while that of 1811 is estimated to have a period of 3,383 years.* The *distances* to which different comets recede from the sun, are also very various. While Encke's comet performs its entire revolution within the orbit of Jupiter, Halley's comet recedes from the sun to twice the

Fig. 78.

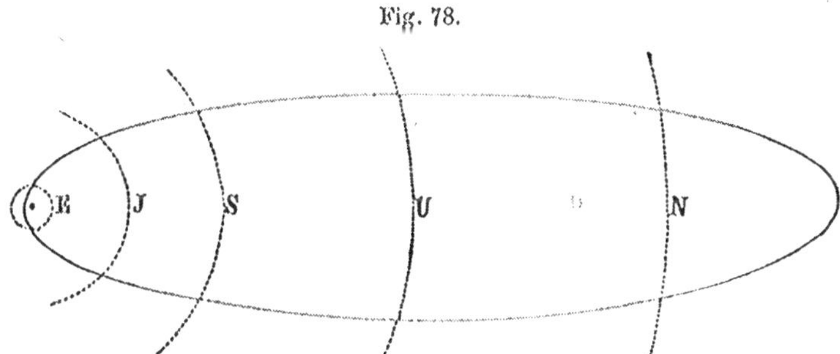

distance of Uranus, or nearly 3,600,000,000 miles. Figure 78 is a representation, in due proportions, of the orbit of this comet. Its vast dimensions will be truly conceived of by reflecting that the radius of the small circle E of the earth's orbit implies a space of nearly 100,000,000 miles; that, as the comet recedes from the sun, it soon reaches the orbit of Jupiter, and successively traverses the orbits of Saturn, Uranus, and Neptune, reaching its aphelion 600,000,000 miles beyond the present boundaries of the planetary system. Some comets, indeed, are thought to go to a much greater distance from the

* Milne.

sun than this, as that of 1811 must have receded from it more than 45,000,000,000 miles, while some even are supposed to pass into parabolic or hyperbolic orbits, and never to return.

394. Comets shine *by reflecting the light of the sun.* In one or two instances they have exhibited distinct *phases*,* although the nebulous matter with which the nucleus is surrounded, would commonly prevent such phases from being distinctly visible, even when they would otherwise be apparent. Moreover, certain qualities of polarized light enable the optician to decide whether the light of a given body is direct or reflected; and M. Arago, of Paris, by experiments of this kind on the light of the comet of 1819, ascertained it to be reflected light.† The tail of a comet usually increases very much as it approaches the sun; and frequently does not reach its maximum until after the perihelion passage. In receding from the sun the tail again contracts, and nearly or quite disappears before the body of the comet is entirely out of sight. The tail is frequently divided into two portions, the central parts, in the direction of the axis, being less bright than the marginal parts. In 1744, a comet appeared which had six tails, spread out like a fan.

The tails of comets extend in a direct line from the sun, although they are usually more or less curved, like a long quill or feather, being convex on the side next to the direction in which they are moving (Fig. 77); a figure which may result from the less velocity of the portions most remote from the sun. Expansions of the envelope have also been at times observed on the side next the sun,‡ but these seldom attain any considerable length.

395. The *quantity of matter* in comets is exceedingly small. Their tails consist of matter of such tenuity that the smallest stars are visible through them. They can only be regarded as great masses of thin vapor, susceptible of being penetrated through their whole substance by the sunbeams, and reflecting them alike from their interior parts and from their surfaces.

* Delambre, t. iii, p. 400. † Francœur, p. 181.

‡ See Dr. Joslin's remarks on Halley's comet, Amer. Journ. Science, xxxi.

It appears, perhaps, incredible that so thin a substance should be visible by reflected light, and some astronomers have held that the matter of comets is self-luminous; but it requires but very little light to render an object visible in the night, and a light vapor may be visible when illuminated throughout an immense stratum, which could not be seen if spread over the face of the sky like a thin cloud. The highest clouds that float in our atmosphere, must be looked upon as dense and massive bodies, compared with the filmy and all but spiritual texture of a comet.* The small quantity of matter in comets is further proved by the fact that *they have sometimes passed very near to some of the planets without disturbing their motions in any appreciable degree.* Thus the comet of 1770, in its way to the sun, got entangled among the satellites of Jupiter, and remained near them four months, yet it did not perceptibly change their motions. The same comet also came very near the earth; so near, that, had its mass been equal to that of the earth, it would have caused the earth to revolve in an orbit so much larger than at present, as to have increased the length of the year 2h. 47m.† Yet it produced no sensible effect on the length of the year, and therefore its mass, as is shown by Laplace, could not have exceeded $\frac{1}{5000}$ of that of the earth, and might have been less than this to any extent. It may indeed be asked, what proof we have that comets have any matter. The answer is, first, they reflect light; second, though not sufficient to disturb so heavy bodies as planets or satellites, yet they are themselves exceedingly disturbed by the action of the planets, and in exact conformity with the laws of universal gravitation. A delicate compass may be greatly agitated by the vicinity of a mass of iron, while the iron is not sensibly affected by the attraction of the needle.

396. By approaching very near to a large planet, a comet may have its orbit entirely changed. This fact is strikingly exemplified in the history of the comet of 1770. At its appearance in 1770, its orbit was found to be an ellipse, requiring for a complete revolution only 5½ years; and the wonder was, that it had not been seen before, since it was a very large and

* Sir J. Herschel. † Laplace.

bright comet. Astronomers suspected that its path had been changed, and that it had been recently compelled to move in this short ellipse, by the disturbing force of Jupiter and his satellites. The French Institute, therefore, offered a high prize for the most complete investigation of the elements of this comet, taking into account any circumstances which could possibly have produced an alteration in its course. By tracing back its movements for some years previous to 1770, it was found that, at the beginning of 1767, it had entered considerably within the sphere of Jupiter's attraction. Calculating the amount of this attraction from the known proximity of the two bodies, it was found what must have been its orbit previous to the time when it became subject to the disturbing action of Jupiter. The result showed that it then moved in an ellipse of greater extent, having a period of 50 years, and having its *perihelion* instead of its aphelion near Jupiter. It was therefore evident why, as long as it continued to circulate in an orbit so far from the center of the system, it was never visible from the earth. In January, 1767, Jupiter and the comet happened to be very near each other, and as both were moving in the same direction, and nearly in the same plane, they remained in the neighborhood of each other for several months, the planet being between the comet and the sun. The consequence was, that the comet's orbit was changed into a smaller ellipse, in which its revolution was accomplished in 5½ years. But as it was approaching the sun in 1779, it happened again to fall in with Jupiter. It was in the month of June that the attraction of the planet began to have a sensible effect; and it was not until the month of October following that they were finally separated.

At the time of their nearest approach, in August, Jupiter was distant from the comet only $\frac{1}{491}$ of its distance from the sun, and exerted an attraction upon it 225 times greater than that of the sun. By reason of this powerful attraction, Jupiter being further from the sun than the comet, the latter was drawn out into a new orbit, which, even at its perihelion, came no nearer to the sun than the planet Ceres. In this third orbit, the comet requires about 20 years to accomplish its revolution; and being at so great a distance from the earth, it is invisible, and will forever remain so, unless, in the course of ages, it shall undergo new perturbations and move again in some smaller orbit as before.*

ORBITS AND MOTIONS OF COMETS.

397. The planets, as we have seen (with the exception of the asteroids, which seem to be an intermediate class of bodies between planets and comets), move in orbits which are nearly circular, and all very near to the plane of the ecliptic, and all move in the same direction from west to east. But the orbits of comets are far more eccentric than those of the planets; they are inclined to the ecliptic at various angles, being sometimes even nearly perpendicular to it; and the motions of comets are sometimes retrograde.

398. The elements of a comet's orbit as usually obtained, are five: (1), *longitude of the node;* (2), *inclination to the ecliptic;* (3), *perihelion distance;* (4), *longitude of perihelion;* (5), *time of perihelion passage.*

In comparing these with the elements of a planetary orbit (Art. 367), we perceive two to be omitted—the *periodic time*, and the *eccentricity;* while *perihelion* distance is substituted for *mean* distance. The reason for this is, that in these orbits, no reliance can be placed on the determination of the size and form of orbit, or the time of describing it, from observations which are limited to a short arc near the perihelion. The complete problem is not only extremely difficult, as Newton pronounced it, but rather wholly impracticable. It is during only a very small portion of their course that they are visible from the earth, and the observations made in that short period can not afterward be verified on more convenient occasions; whereas in the case of the planets, whose orbits are nearly circular, and whose movements may be followed uninterruptedly throughout a complete revolution, no such impediments to the determination of their orbits occur. There is also some unavoidable uncertainty in observations made upon bodies whose outlines are so ill-defined. It is not unfrequently the case too, that comets move in a direction opposite to the order of the

* Milne.

signs in the zodiac, and sometimes nearly perpendicular to the plane of the ecliptic; so that their apparent course through the heavens is rendered extremely complicated, on account of the contrary motion of the earth. Since it is possible there should be any number of elliptic orbits, whose perihelion distances are equal, it is obvious that, in the case of very eccentric orbits, the slightest change in the position of the curve near the vertex, where alone the comet can be observed, must occasion a very sensible difference in the length of the orbit (as will be obvious from Fig. 79); and therefore, though a small error

Fig. 79.

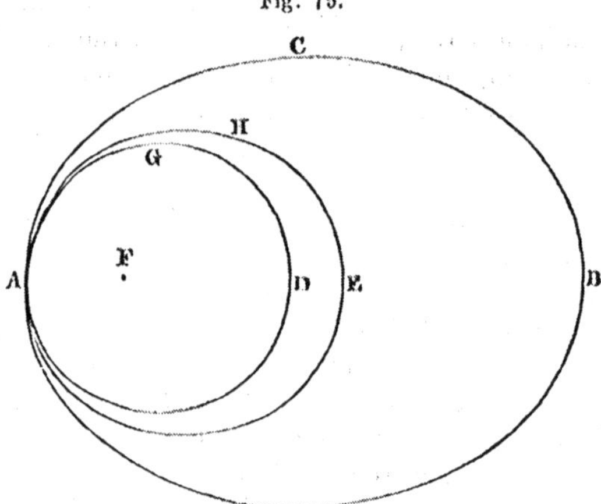

produces no perceptible discrepancy between the observed and the calculated course, while the comet remains visible from the earth, its effect, when diffused over the whole extent of the orbit, may acquire a most material or even a fatal importance.

On account of these circumstances, it is found exceedingly difficult to lay down, even in the rudest manner, the path which a comet actually pursues, when gone for years beyond the limit of our vision; and least of all to determine with accuracy the length of the major axis of the ellipse, so as to derive from it, by Kepler's third law, the time of its revolution, and thus to be able to predict its next perihelion passage. An error of only a few seconds may cause a difference of many hundred years. In this manner, though Bessel determined the revolution of the comet of 1769 to be 2,089 years, it was found that an error of no more than 5″ in observation, would alter the period either to 2,678 years, or to 1,692 years. Some astronomers, in calculating the orbit of the great comet of 1680, have found the length of its greater axis 426 times the earth's distance from the sun, and consequently its period 8,792 years; while others estimate the greater axis 430 times the earth's distance, which alters the period to 8,916 years. Newton and Halley, however, judged that this comet accomplished its revolution in only 570 years.

399. Disheartened by the difficulty of attaining to any precision in the circumstances by which an elliptic orbit is characterized, and, moreover, taking into account the laborious calculations necessary for its investigation, astronomers usually satisfy themselves with ascertaining the elements of a comet on the supposition of its describing a parabola; and, as this is a curve whose axis is infinite, the procedure is greatly simplified by leaving entirely out of consideration the periodic revolution. It is true that a parabola may not represent, with mathematical strictness, the course which a comet actually follows; but as a parabola is the intermediate curve between the hyperbola and ellipse, it is found that this method, which is so much more convenient for computation, also accords sufficiently with observations, except in cases when the ellipse is a comparatively short one, as that of Encke's comet, for example. When the elements of a comet are determined, Kepler's law of areas enables astronomers to find, by *computation*, the exact place of the comet in its orbit at any given time, on the supposition that its path is a parabola; and comparing this place with that determined by *observation* for the same instant, it is seen whether the orbit is truly parabolic, or whether its deviation from that path is such as to indicate that its real path is an ellipse; and the amount of such deviation will give some idea of the degree of eccentricity of the ellipse.

400. The elements of a comet, with the exception of its periodic time, are calculated in a manner similar to those of the planets. Three good observations on the right ascension and declination of the comet (which are usually found by ascertaining its position with respect to certain stars, whose right ascensions and declinations are accurately known) afford the means of calculating these elements.

The appearances of the same comet at different periods of its return are so various (Art. 392), that we can never pronounce a given comet to be the same with one that has appeared before, from any peculiarities in its physical aspect. The identity of a comet with one already on record, is determined by the *identity of the elements.* When a new comet appears, we first determine its elements, and then turning to a catalogue of comets whose elements have previously been found and placed on record, we see whether these new elements agree with any set of those in the catalogue. If they do, we infer that the present comet is identical with that on record; and the interval between the two appearances of the body will indicate its periodic time. If, for example, we find respecting a comet now visible in the sky, that its path makes the same angle with the ecliptic as that of a certain comet in our catalogue, that it crosses the ecliptic in the same degree of longitude, that it comes to its perihelion in the same place, that its perihelion distance is the same, and its course the same in regard to the order of the signs, then we infer that the two bodies are one and the same; and the number of years that have elapsed since its former appearance indicates the period of its revolution around the sun. But if these particulars differ wholly from any set of recorded elements, we infer that the present is a comet which has never visited our sphere before, or at least one whose elements have not been determined and recorded. It was by this means that Halley first established the identity of the comet which bears his name, with one that had appeared at several preceding ages of the world, of which so many particulars were left on record, as to enable him to calculate the elements at each period. These were as in the following table.

Time of appearance.	Inclination of the Orbit.	Longitude of the Node.	Longitude of Perihelion.	Perihelion Distance.	Course.
1456	17° 56′	48° 30′	301° 00	0.58	Retrograde.
1531	17 56	49 25	301 39	0.57	Retrograde.
1607	17 02	50 21	302 16	0.58	Retrograde.
1682	17 42	50 48	301 36	0.58	Retrograde.

On comparing these elements, no doubt could be entertained that they belonged to one and the same body; and since the interval between the successive returns was seen to be 75 or 76 years, Halley ventured to predict that it would again return in 1758. Accordingly, the astronomers who lived at that period looked for its return with the greatest interest. It was found, however, that on its way toward the sun it would pass very near to Jupiter and Saturn, and by their action on it, would be retarded for a long time. Clairaut, a distinguished French mathematician, undertook the laborious task of estimating the exact amount of this retardation, and found it to be no less than 618 days, namely, 100 days by the action of Jupiter, and 518 days by that of Saturn. This would delay its appearance until early in the year 1759, and Clairaut fixed its arrival at the perihelion within a month of April 13th. It came to the perihelion on the 12th of March.

101. The return of Halley's comet in 1835 was looked for with no less interest than in 1759. Several of the most accurate mathematicians of the age had calculated its elements with inconceivable labor. Their zeal was rewarded by the appearance of the expected visitant at the time and place assigned; it traversed the northern sky, presenting the very appearances, in most respects, that had been anticipated; and came to its perihelion on the 16th of November, within one day of the time predicted by Pontecoulant, a French mathematician, who had, it appeared, made the most successful calculation.* On its previous return, it was deemed an extraordinary achievement to have brought the prediction within a month of the actual time.

Many circumstances conspired to render this return of Halley's comet an astronomical event of transcendent interest. Of all the celestial bodies, its history was the most remarkable; it afforded most triumphant evidence of the truth of the doctrine of universal gravitation, and consequently of the received laws of astronomy; and it inspired new confidence in the power of

* See Professor Loomis's Observations on Halley's Comet, Amer. Journ. Science, xxx., 209. Pontecoulant's Phys. Céleste Précis, p. 586.

that instrument (the Calculus) by means of which its elements had been investigated.

402. Encke's comet, by its frequent returns, affords peculiar facilities for ascertaining the laws of its revolution; and it has kept the appointments made for it with great exactness. On its return in 1839, it exhibited to the telescope a globular mass of nebulous matter resembling fog, and moved toward its perihelion with great rapidity.

But what has made Encke's comet particularly famous, is its having first revealed to us the existence of a *Resisting Medium* in the planetary spaces. It has long been a question whether the earth and planets revolve in a perfect void, or whether a fluid of extreme rarity may not be diffused through space. A perfect vacuum was deemed most probable, because no such effects on the motions of the planets could be detected as indicated that they encountered a resisting medium. But a feather or a lock of cotton, propelled with great velocity, might render obvious the resistance of a medium which would not be perceptible in the motions of a cannon-ball. Accordingly, Encke's comet is thought to have plainly suffered a retardation from encountering a resisting medium in the planetary regions. The effect of this resistance, from the first discovery of the comet to the present time, has been to diminish the time of its revolution about two days. Such a resistance, by destroying part of the projectile force, would cause the comet to approach nearer to the sun, and thus to have its periodic time shortened. The ultimate effect of this cause will be to bring the comet nearer to the sun at every revolution, until it finally falls into that luminary, although many thousand years will be required to produce this catastrophe.* It is conceivable, indeed, that the effects of such a resistance may be counteracted by the attraction of one or more of the planets near which it may pass in its successive returns to the sun. It is peculiarly interesting to see a portion of matter of a tenuity exceeding the thinnest fog, pursuing its path in space, in obedience to the same laws as those which regulate such large and heavy bodies as Jupiter

* Halley's comet, at its return in 1835, did not appear to be affected by the supposed resisting medium, and its existence is considered as still doubtful.

or Saturn. In a perfect void, a speck of fog, if propelled by a suitable projectile force, would revolve around the sun, and hold on its way through the widest orbit, with as sure and steady a pace as the heaviest and largest body in the system.

403. The most remarkable comet of the present century hitherto observed, was the great comet of 1843. (See Plate I. at the end of the volume.) On the 28th of February of that year, the attention of numerous observers in various parts of the world was arrested by a comet seen in the broad light of day, a little eastward of the sun. In Mexico it was observed, and its altitude repeatedly measured with a sextant, from nine in the morning until sunset. In New England, it was seen at several places from half-past seven in the morning until after three in the afternoon, when the sky became obscured by haziness and clouds. Accurate measures were taken by Capt. Clark, at Portland, Maine, of the distance of the nucleus from the sun's limb. At 3h. 2m. 15s. mean time, the distance of the sun's furthest limb from the nearest limb of the nucleus was 4° 6′ 15″. The comet resembled a white cloud of great density, being nearly equally shining throughout, with a nucleus as bright as the full moon at midnight in a clear sky. During the first week in March, the appearance of this body was splendid and magnificent, enhanced in both respects by the transparency of a tropical sky, and the higher angle of elevation above that at which it was seen by northern observers. At New Haven, it was first seen after sunset, on the 5th of March. It then lay far in the southwest. On account of the presence of the moon, it was not seen most favorably until the evening of the 17th. It then extended along the constellation Eridanus to the ears of the Hare, below the feet of Orion, reaching nearly to Sirius, being about 40° in length, although in the tropical regions its apparent length, at its maximum, was nearly 70°. It was slightly curved like a goose-quill, and colored with a slight tinge of rose-red, which in a few evenings disappeared, and left it nearly a pearly white. Our diagram (Plate I. at the end of the volume) presents a pretty accurate idea of its appearance on the 20th of March. All the astronomers of the age have agreed in the opinion that this is one of the most remarkable exhibitions of a comet ever witnessed, although they are not fully agreed respecting the elements of its orbit, or its periodic time. Its elements resemble those of the comet of 1688, which would give a period of 175 years; and to this periodic time, authority at present inclines;* but Prof. Hubbard, of the Washington Observatory, after an elaborate discussion of all the observations, thinks the most probable period 170 years.

Of all the comets on record, the great comet of 1843 approached nearest to the sun. It came within about 60,000 miles of his luminous surface, or only about one-fourth of the distance of the moon from the earth. It will be recollected that to a spectator on the earth the sun's angular diameter but a little exceeds half a degree; but were we to approach as near to the sun as this body did in its perihelion, that diameter would appear no less than 121° 32′; and the light and heat (which increase as the square of the distance is diminished) would be 47,000 times as great as at present, the heat exceeding nearly twenty-five times that produced by Parker's great burning lens, although this instrument is capable of producing effects beyond those of the most powerful blast-furnace. The velocity of the comet was still more astonishing, being at the rate of more than one and a quarter million of miles per hour,—a velocity sufficient to carry it through 180°, or half round the sun, in two hours.†

An interesting comet appeared in 1858, called the comet of Donati, who first saw it at Florence, June 2d. It continued in sight till Oct. 15th. Its tail, when at perihelion, Oct. 10th, was 60° long. Its nucleus was uncommonly bright, and beyond it, in the axis of the envelope and tail was a dark, straight line, like a shadow. The long period of its visibility gave unusual opportunity for careful observations. Its periodic time is variously estimated, from 1,620 to 2,495 years.‡

404. Of the *physical nature* of comets, little is understood. It is usual to account for the variations which their tails un-

* See American Almanac for 1844, p. 94. Amer. Journal of Science, xlv., p. 188. Astronomical Journal, vol. ii., p. 156.

† Herschel's Outlines, p. 318.

‡ See Bond's "Account of Donati's Comet," with fine illustrations, in Math. Monthly, Dec., 1853.

dergo by referring them to the agencies of heat and cold. The intense heat to which they are subject in approaching so near the sun as some of them do, is alleged as a sufficient reason for the great expansion of the thin nebulous atmospheres forming their tails; and the inconceivable cold to which they are subject in receding to such a distance from the sun, is supposed to account for the condensation of the same matter until it returns to its original dimensions. The temperature experienced by the comets of 1680 and 1843 at their perihelion, would be sufficient to volatilize the most obdurate substances, and to expand the vapor to vast dimensions; and the opposite effects of the extreme cold to which it would be subject in the regions remote from the sun, would be adequate to condense it into its former volume.

This explanation, however, does not account for the direction of the tail, extending, as it usually does, only in a line opposite to the sun. Some writers, therefore, as Delambre,* suppose that the nebulous matter of the comet, after being expanded to such a volume that the particles are no longer attracted to the nucleus unless by the slightest conceivable force, is carried off in a direction from the sun by the impulse of the solar rays themselves. Others conceive of a force of *repulsion*, independent of any mechanical impulse emanating from the sun. But to assign such a power of communicating motion to the sun's rays while they have never been proved to have any momentum, or to a repulsive force which has no independent proof of its existence, is unphilosophical; and we are compelled to place the phenomena of comets' tails among the points of astronomy yet to be explained.†

405. Since those comets which have their perihelion very near the sun, like the comet of 1680, cross the orbits of all the planets, *the possibility that one of them may strike the earth*, has frequently been suggested. Still, it may quiet our apprehensions on this subject, to reflect on the vast extent of the planetary spaces, in which these bodies are not crowded together as

* Delambre's Astronomy, t. iii., p. 401.

† Prof. Norton on the "Formation of Comets' Tails," (Amer Journal, xlvii., p. 104).

we see them erroneously represented in orreries and diagrams, but are sparsely scattered at immense distances from each other. They are like insects flying in the expanse of heaven. If a comet's tail lay with its axis in the plane of the ecliptic when it was near the sun, we can imagine that the tail might sweep over the earth; but the tail may be situated at any angle with the ecliptic as well as in the same plane with it, and the chances that it will not be in the same plane are almost infinite. It is also extremely improbable that a comet will cross the plane of the ecliptic precisely at the earth's path in that plane, since it may as probably cross it at any other point nearer or more remote from the sun. Still, some comets have occasionally approached near to the earth. Thus Biela's comet, in returning to the sun in 1832, crossed the ecliptic very near to the earth's track, and had the earth been then at that point of its orbit, it might have passed through a portion of the nebulous atmosphere of the comet. The earth was within a month of reaching that point. This might at first view seem to involve some hazard; yet we must consider that a month short implied a distance of nearly 50,000,000 miles. Laplace has assigned the consequences that would ensue in case of a direct collision between the earth and a comet;* but terrible as he has represented them on the supposition that the nucleus of the comet is a solid body, yet considering a comet (as most of them doubtless are) as a mass of exceedingly light nebulous matter, it is not probable, even were the earth to make its way directly through a comet, that a particle of the comet would reach the earth. The portions encountered by the earth, would be arrested by the atmosphere, and probably inflamed; and they would perhaps exhibit, on a more magnificent scale than was ever before observed, the phenomena of shooting stars or meteoric showers.

METEORIC SHOWERS.

406. The remarkable exhibitions of shooting stars which have occurred within a few years past, have excited great interest among astronomers, and led to some new views respecting the construction of the solar system. Their attention was

* Syst. du Monde, l. iv., c. 4.

first turned toward this subject by the great meteoric shower of November 13, 1833. On that morning, from two o'clock until broad daylight, the sky being perfectly serene and cloudless, the whole heavens were lighted with a magnificent display of celestial fireworks. Numerous bright bodies, which might be compared with stars of the largest magnitudes, and with planets, were darting toward the earth on all sides, describing arcs of great circles, of all lengths from 70° to less than a single degree. In many cases, they left long trains of light in their paths, which lasted a few seconds; and occasionally, when a meteor of unusual brightness descended, the train of light continued for minutes. The light which some of them shed was equal to that of the moon at the quarter. The whole number seen at any one place of observation could not have been less than 200,000.

On tracing back the lines of direction in which the meteors moved, it was found that they all appeared to radiate from the same point, which was situated near one of the stars (*Gamma Leonis*) of the sickle, in the constellation Leo; and, in every repetition of the meteoric shower of November, the radiant point has occupied nearly the same situation.

This shower pervaded nearly the whole of North America, having appeared in almost equal splendor from the British possessions on the north, to the West India Islands and Mexico on the south, and from sixty-one degrees of longitude east of the American coast, quite to the Pacific ocean on the west. Throughout this immense region the duration was nearly the same. The meteors began to attract attention by their unusual frequency and brilliancy, from *nine to twelve* o'clock in the evening; were most striking in their appearance from *two to four;* arrived at their maximum, in many places, about *four* o'clock; and continued until rendered invisible by the light of day. The meteors moved in right lines, or in such apparent curves, as, upon optical principles, can be resolved into right lines. Their general tendency was toward the northwest, although by the effect of perspective they appeared to move in all directions.

407. Soon after this occurrence, it was ascertained that a similar meteoric shower had appeared in 1799, and what was remarkable, almost exactly at the same time of the year, namely, on the morning of the 12th of November; and it soon appeared, by accounts received from different parts of the world, that this phenomenon had occurred on the same 13th of November, in 1830, 1831, and 1832. Hence, this was evidently an event independent of the casual changes of the atmosphere; for, having a periodical return, it was undoubtedly to be referred to astronomical causes, and its recurrence, at a certain definite period of the year, plainly indicated *some* relation to the revolution of the earth around the sun.

It remained, however, to develop the nature of this relation, by investigating, if possible, the origin of the meteors. The views to which the author of this work was led, suggested the probability that the same phenomenon would recur at the corresponding seasons of the year for at least several years afterward; and such proved to be the fact, although the appearances, at every succeeding return, were less and less striking, until 1839, when, so far as is known, they ceased altogether.

Meanwhile, three other distinct periods of meteoric showers have also been determined; one on the 9th of August, and (more rare) on the 21st of April and 7th of December respectively.

408. The following *conclusions* respecting the meteoric shower of November, are believed to be well established, and most of them are now generally admitted by astronomers, though we can not here exhibit the evidence on which they were founded.*

It is considered, then, as established, that the periodical meteors of November (and most of the conclusions apply equally to those of August) have their *origin* beyond the atmosphere, descending to us from some body (which, from the known constitution of the meteors, may be called a *nebulous* body) with which the earth falls in, and near or through the borders of which it passes; that this body has an independent existence as a member of the solar system, its periodic

* We beg leave to refer the reader to various publications on the subject, by the author and others, in the American Journal of Science, commencing with the 25th volume; and also to "Letters on Astronomy," by the author of this work.

time being nearly commensurable with the earth's, either a year or half a year, so that for a number of years in succession the two bodies meet near the same part of the earth's orbit. It is further established, that the meteors consist of light combustible matter; that they move with great velocities, amounting, in some instances, to not less than that of the earth in its orbit, or 19 miles per second; that some of them are bodies of large size, sometimes several thousand feet in diameter; that when they enter the atmosphere, they rapidly and powerfully condense the air before them, and thus elicit the heat that sets them on fire, as a spark is elicited in the air-match, by being suddenly condensed by means of a piston and cylinder; and that they are burned up at a considerable height above the earth, sometimes not less than 30 miles.

409. Calling the body from which the meteors descended the "meteoric body," it is inferred that it is a body of great extent, since, without apparent exhaustion, it has been able to afford such copious showers of meteors at so many different times; and hence we regard the part that has descended to the earth only as the *extreme portions* of a body or collection of meteors, of unknown extent, existing in the planetary spaces. Since the earth fell in with the meteoric body, in the same part of its orbit for several years in succession, the body must either have remained there while the earth was performing its whole revolution around the sun, or it must itself have had a revolution, as well as the earth. No body can remain stationary within the planetary spaces; for, unless attracted to some nearer body, it would be drawn directly toward the sun, and could not have been encountered by the earth again in the same part of her orbit. Nor can any mode be conceived in which this event could have happened so many times in regular succession, unless the body had a revolution of its own around the sun. Finally, to have come into contact with the earth at the same part of her orbit, in two or more successive years, the body must have a period which is either nearly the same with the earth's period, or some aliquot part of it. No period will fulfil these conditions, but either a year or half a year. Which of these is the true period of the meteoric body, is not fully determined.

PART III.—OF THE FIXED STARS AND SYSTEM OF THE WORLD.

CHAPTER I.

OF THE FIXED STARS—CONSTELLATIONS.

410. The Fixed Stars are so called, because, to common observation, they always maintain the same situations with respect to one another.

The stars are classed by their apparent *magnitudes*. The whole number of magnitudes recorded are *sixteen*, of which the first six only are visible to the naked eye; the rest are *telescopic stars*. As the stars which are now grouped together under one of the first six magnitudes are very unequal among themselves, it has recently been proposed to subdivide each class into three, making in all eighteen instead of six magnitudes visible to the naked eye. These magnitudes are not determined by any very definite scale, but are merely ranked according to their relative degrees of brightness, and this is left in a great measure to the decision of the eye alone, although it would appear easy to measure the comparative degree of light in a star by a photometer, and upon such measurement to ground a more scientific classification of the stars. The brightest stars to the number of 15 or 20 are considered as stars of the *first* magnitude; the 50 or 60 next brightest, of the *second* magnitude; the next 200 of the *third* magnitude; and thus the number of each class increases rapidly as we descend the scale, so that no less than fifteen or twenty thousand are included within the first seven magnitudes.

411. The stars have been grouped in *Constellations* from the most remote antiquity; a few, as Orion, Bootes, and Ursa Major, are mentioned in the most ancient writings under the same names as they bear at present. The names of the constellations are sometimes founded on a supposed resemblance to the objects to which the names belong; as the Swan and the Scorpion were evidently so denominated from their likeness to those animals; but in most cases it is impossible for us to find any reason for designating a constellation by the figure of the animal or the hero which is employed to represent it. These representations were probably once blended with the fables of pagan mythology. The same figures, absurd as they appear, are still retained for the convenience of reference; since it is easy to find any particular star, by specifying the part of the figure to which it belongs, as when we say a star is in the neck of Taurus, in the knee of Hercules, or in the tail of the Great Bear. This method furnishes a general clue to its position; but the stars belonging to any constellation are distinguished according to their apparent magnitudes, as follows: first, by the Greek letters, Alpha, Beta, Gamma, &c. Thus α *Orionis*, denotes the largest star in Orion, β *Andromedæ*, the second star in Andromeda, and γ *Leonis*, the third brightest star in the Lion. Where the number of the Greek letters is insufficient to include all the stars in a constellation, recourse is had to the letters of the Roman alphabet, a, b, c, &c.; and in cases where these are exhausted, the final resort is to numbers. This is evidently necessary, since the largest constellations contain many hundreds or even thousands of stars. *Catalogues* of particular stars have also been published by different astronomers, each author numbering the individual stars embraced in his list, according to the places they respectively occupy in the catalogue. These references to particular catalogues are sometimes entered on large celestial globes. Thus we meet with a star marked 84 H., meaning that this is its number in Herschel's catalogue, or 140 M., denoting the place the star occupies in the catalogue of Mayer.

412. The earliest catalogue of the stars was made by Hipparchus, of the Alexandrian school, about 140 years before the Christian era. A new star appearing in the firmament, he was induced to count the stars and to record their positions, in order that posterity might be able to judge of the permanency of the constellations. His catalogue contains all that were *conspicuous* to the naked eye in the latitude of Alexandria, being 1,022.

Most persons unacquainted with the actual number of the stars which compose the visible firmament, would suppose it to be much greater than this; but it is found that the catalogue of Hipparchus embraces nearly all that are easily seen in the same latitude, and that on the equator, where the spectator has the northern and southern hemispheres both in view, the number of stars that can be counted does not exceed 3000. A hasty glance over the sky gives us the impression of a countless multitude of stars; but the greater part vanish as soon as we try to number them. This is owing to the *indirect* vision of thousands of faint stars, which are unseen as soon as we turn the axes of the eyes directly upon them.

By the aid of the telescope, new fields of stars present themselves of boundless extent; the number continually augmenting as the powers of the telescope are increased. Lalande, in his Histoire Céleste, has registered the positions of no less than 50,000; and the whole number visible in the largest telescopes amount to many millions.

413. It is strongly recommended to the learner to acquaint himself with the leading constellations at least, and with a few of the most remarkable individual stars. The task of learning them is comparatively easy, when they are taken up at suitable intervals throughout the year, the moon being absent and the sky clear. After becoming familiar with such constellations as are visible on any given evening (suppose the first of January), these may be carefully reviewed after an interval of a month, and the several new ones added which have in the mean time risen above the eastern horizon. By repeating this process near the beginning of every month of the year, the learner will acquire a competent knowledge of the whole that are visible in his latitude, and with a small expenditure of time. It may at first be advisable to obtain, for an evening or two, the assistance of some one who is acquainted with the constellations, to point out such as are then visible in the evening sky. Then, by the aid of a celestial map, or, what is better, a celestial globe, the learner will pursue the study without difficulty. We begin by rectifying the globe for the time, according to the directions given in Article 76.

In the following sketch of the leading constellations, we will point out a few of the marks by which they may be severally recognized, adding occasionally a few particulars, and leaving it to the learner to fill up the outline by the aid of his map or globe, one of which, indeed, is presumed to be before him.*

Let us begin with the constellations of the *Zodiac*, which, succeeding each other as they do in a known order, are most easily found.†

ARIES (the RAM), the first constellation of the Zodiac, is known by two bright stars, *Alpha* (α) on the northeast, and *Beta* (β) on the southwest, 4°‡ apart, forming the head. South of Beta, at the distance of 2°, is a smaller star, *Gamma* (γ). The next brightest star of the Ram, *Delta* (δ), is in the tail, 15° southeast of Alpha. The feet of the figure rest on the head of the Whale. It has been already intimated (Art. 193), that the vernal equinox was near the head of Aries, when the signs of the Zodiac received their present names, but that the equinox is now found 30° westward of α Arietis, in consequence of the precession of the equinoxes.

TAURUS (the BULL) will be readily found by the seven stars, or *Pleiades*, which lie in the neck, 24° eastward of α Arietis. The largest star in Taurus is *Aldebaran*, of the first magnitude, in the Bull's eye, 14° southeast of the Pleiades. It has a reddish color, and resembles the planet Mars. The other eye of

* A celestial globe, sufficient for studying the constellations, may be purchased for a small sum, and is, in other respects, a valuable possession to the astronomical student; but even cheap maps of the stars, like those of Burritt or Kendal, will answer for beginners; and the Celestial Atlas, published by the Society for the Diffusion of Useful Knowledge, which is suitable for the more advanced student, may be procured at a moderate expense.

† It will be expedient, where it is practicable, for the learner to study the constellations in separate portions, at different seasons of the year, as at the equinoxes and at the solstices, according to the directions given in the closing article of this chapter.

‡ These measures are not intended to be stated with minute accuracy, but only with such a degree of exactness as may serve for a general guide. The learner will find it greatly for his advantage to accustom himself to make an accurate estimate with the eye of distances in degrees on the celestial sphere; and he may, at the outset, fix on the distance between Alpha and Beta Arietis as a standard measure (4°) by which to estimate other angular distances among the stars. Thus, half this length applied from Beta to Gamma, indicates that the two latter stars are 2° apart; and two and a half times the same measure (10°) will reach from the Pleiades to Aldebaran. Or the Pointers in the Great Bear will furnish a measure of 5°.

the figure is *Epsilon* (ϵ), 3° northwest of Aldebaran. Five small stars, situated a little west of Aldebaran, in the face of the Bull, constitute the *Hyades*. Although the Pleiades are usually denominated the *seven* stars, yet it has been remarked, from a high antiquity, that only six are present.

Quæ septem dici, sex tamen esse solent.*—*Ovid.*

Some persons, however, of remarkable powers of vision, are still able to recognize seven, and even a greater number.† With a moderate telescope, not less than 50 or 60 stars, of considerable brightness, may be counted in this group, and a much larger number of very small stars are revealed to the more powerful telescopes. The beautiful allusion, in the book of Job, to the "sweet influences of the Pleiades," and the special mention made of this group by Homer and Hesiod, show how early it had attracted the attention of mankind. The *horns* of the Bull are two stars, *Beta* and *Zeta*, situated 25° east of the Pleiades, being 8° apart. The northern horn, *Beta*, also forms one of the feet of *Auriga*, the Charioteer.

GEMINI (the TWINS) is represented by two well-known stars, Castor and Pollux, in the head of the figure, 5° asunder. Castor, the northern, is of the first, and Pollux of the second magnitude. Four conspicuous stars, extending in a line from south to north, 25° S. W. of Castor, form the feet, and two others, parallel to these at the distance of six or seven degrees northeastward, are in the knees.

CANCER (the CRAB). There are no large stars in this constellation, and it is regarded as less remarkable than any other in the Zodiac. The two most conspicuous stars, *Alpha* and *Beta*, are in the southern claws of the figure, and in its body are the northern and southern *Asellus*, which may be readily found on a celestial globe. But the most remarkable object in this constellation is a misty group of very small stars, so close together, when seen by the naked eye, as to resemble a comet,

* Their names were Electra, Maia, Taygeta, *Alcyone*, Celæno, Asterope, and Merope, the last being the "Lost Pleiad" of the poets. Alcyone, according to a recent celebrated hypothesis, is distinguished as the center around which the starry host revolve.

† Smyth's Cycle, ii., p. 86.

but easily separated by the telescope into a beautiful collection of brilliant points. It is called *Præsepe*, or the *Beehive*.

LEO (the LION) is a very large constellation, and has many interesting members. *Regulus* (*α Leonis*) is a star of the first magnitude, which lies very near the ecliptic, and is much used in astronomical observations. North of Regulus lies a semicircle of five bright stars, arranged in the form of a *sickle*, of which Regulus is the handle, and extending over the shoulder and neck of the Lion.* *Denebola*, a conspicuous star in the Lion's tail, lies 25° east of Regulus. Twenty bright stars in all help to compose this beautiful constellation. It ranges from west to east along the Zodiac, over more than 40° of longitude, all parts of the figure excepting the feet lying north of the ecliptic.

VIRGO (the VIRGIN) extends along the Zodiac eastward from the Lion, covering an equally wide region of the heavens, although less distinguished by brilliant stars. *Spica*, however, is a star of the first magnitude, and lies a little east of the vernal equinox. *Vindemiatrix*, in the arm of Virgo, 18° east of Denebola, and 23° north of Spica, is easily found, and directly south of Denebola 13°, is β Virginis; while four other conspicuous stars, in the form of a trapezium, between this and Vindemiatrix, lie in the wing and shoulders of the figure. The feet are near the Balance.

LIBRA (the BALANCE) is composed of a few scattered members situated between the feet of Virgo and the head of Scorpio, but has no very distinctive marks. Two stars of the second magnitude, *Alpha* on the south, and *Beta* 8° northeast of Alpha, together with a few smaller stars, form the scales.

SCORPIO (the SCORPION) is one of the finest of the constellations of the Zodiac, and is manifestly so called from its resemblance to the animal whose name it bears. The head is composed of five stars, arranged in a line slightly curved, which is crossed in the center by the ecliptic, nearly at right angles, a degree south of the brightest of the group β *Scorpionis*. Nine degrees southeast of this is a remarkable star of the first mag-

* As the *Meteors of November* always appear to radiate from a point in the bend of the sickle, near the star Gamma, it may be noted that the names of the six stars composing this figure, beginning with Regulus, are α, η, γ, ζ, μ, ε.

nitude, called *Antares*, and sometimes the *Heart of the Scorpion* (*Cor Scorpionis*). It is of a red color, resembling the planet Mars. South and east of this, a succession of not less than nine bright stars sweep round in a semicircle, terminating in several small stars forming the sting of the Scorpion. The tail of the figure extends into the Milky Way.

Sagittarius (the Archer). Ten degrees eastward of the Scorpion's tail, on the eastern margin of the Milky Way, we come to the bow of Sagittarius, consisting of three stars about 6° apart, the middle one being the brightest, and situated in the bend of the bow, while a fourth star, 4° westward of it, constitutes the arrow. The archer is represented by the figure of a Centaur (half horse and half man), and proceeding about ten degrees east from the bow, we come to a collection of seven or eight stars of the second and third magnitudes, which lie in the human or upper part of the figure.

Capricornus (the Goat), represented with the head of a goat and the tail of a fish, comes next to Sagittarius, about 20° eastward of the group that form the upper portions of that constellation. Two stars of the second magnitude, α on the north, and β on the south, 3° apart, constitute the head of Capricornus, while a collection of stars of the third magnitude, lying 20° southeast of these form the tail.

Aquarius (the Water Bearer) is closely in contact with the tail of Capricornus, immediately north of which, at the distance of 10°, is the western shoulder (β), and 10° further east is the eastern shoulder (α) of Aquarius. About 3° southeast of α is γ Aquarii, which, together with the other two, makes an acute triangle, of which β forms the vertex. In the eastern arm of Aquarius are found four stars, which together make the figure Y, the open part being westward, or toward the shoulders of the constellation. Aquarius ranges nearly 30° from north to south, being nearly bisected by the ecliptic.

Pisces (the Fishes). Three figures of this kind, at a great distance apart, two north and one south of the ecliptic, compose this constellation. The Southern Fish, *Piscis Australis*, otherwise called *Fomalhaut*, lies directly below the feet of Aquarius, and being the only conspicuous star in that part of the heavens, is much used in astronomical measurements. It is 30° south of the equator.

About 12° east of the figure Y in the arm of Aquarius, is an assemblage of five stars, forming a pretty regular *pentagon*, which is one of the northern members of the constellation Pisces; and far to the northeast of this figure, north of the head of Aries, lies the third member, the three being represented as connected together by a *ribbon*, or wavy band, composed of minute stars.

414. The Constellations of the Zodiac being first well learned, so as to be readily recognized, will facilitate the learning of others that lie north and south of them. Let us therefore next review the principal *Northern Constellations*, beginning at the North Pole.

Ursa Minor (the Little Bear). The *Pole-star* (*Polaris*) is in the extremity of the tail of the Little Bear. It is of the third magnitude, and being within less than a degree and a half of the North Pole of the heavens, it serves at present to indicate the position of the pole. It will be recollected, however, that on account of the precession of the equinoxes, the pole of the heavens is constantly shifting its place from east to west, revolving about the pole of the ecliptic, and will in time recede so far from the pole-star, that this will no longer retain its present distinction (Art. 190). Three stars in a straight line, 4° or 5° apart, commencing with Polaris, lead to a trapezium of four stars, the whole seven together forming the figure of a *dipper*, the trapezium being the body, and the three first-mentioned stars being the handle.

Ursa Major (the Great Bear) is one of the largest and most celebrated of the constellations. It is usually recognized by the figure of a larger and more perfect *dipper* than the one in the Little Bear—three stars, as before, constituting the handle, and four others, in the form of a trapezium, the body of the figure. The two western stars of the trapezium, ranging nearly with the North Star, are called the *Pointers*; and beginning with the northern of these two, and following round from left to right through the whole seven, they correspond in rank to the succession of the first seven letters of the Greek alphabet, *Alpha*, *Beta*, *Gamma*, *Delta*, *Epsilon*, *Zeta*, *Eta*. Several of them also are known by their Arabic names. Thus, the first in the tail, corresponding to Epsilon, is *Alioth*, the next (Zeta) *Mizar*, and the last (Eta) *Benetnasch*. These are all bright and beautiful stars, Alpha being of the first magnitude; Beta, Gamma, Delta, of the second; and the three forming the tail, of the third. But it must be remarked that this very remarkable figure of a dipper or ladle composes but a small part of the entire constellation, being merely the hinder half of the body and the tail of the Bear. The head and breast of the figure, lying about ten or twelve degrees west of the Pointers, contain a great number of minute stars in a triangular group. One of the fourth magnitude, *Omicron*, is in the mouth of the Bear. The feet of the figure may be looked for about 15° south of those already described, the two hinder paws consisting each of two stars very similar in appearance, and only a degree and a half apart. The two paws are distant from each other about 18°; and following westward about the same number of degrees, we come to another very similar pair of stars, which constitute one of the fore paws, the other foot being without any corresponding pair.

In a clear winter's night, when the whole constellation is above the pole, these various parts may be easily recognized, and the entire figure will be seen to resemble a large animal, readily accounting for the name given to this constellation from the earliest ages.

DRACO (the DRAGON) is also a very large constellation, extending for a great length from east to west. Beginning at the tail which lies half way between the Pointers and the Pole-star, and winding round between the Great and the Little Bear, by a continued succession of bright stars from 5° to 10° asunder, it coils around under the feet of the Little Bear, sweeps round the pole of the ecliptic, and terminates in a trapezium formed by four conspicuous stars, from thirty to thirty-five degrees from the North Pole. A few of the members of this constellation are of the second, but the greater part of the third magnitude, and below it.

415. With the constellations already described as general landmarks, we may now proceed with each of the principal remaining ones, by stating its *boundaries*, as we do those of countries in geography; their relative situations being thus first learned from a map, or (what is better) from a celestial globe, and then being severally traced out on the sky itself. We will begin with those which surround the North Pole.

CEPHEUS (the KING) is bounded N. by the Little Bear, E. by Cassiopeia, S. by the Lizard, and W. by the Dragon. The head lies in the Milky Way, and the feet extend toward the pole. It contains no stars above the third magnitude.

CASSIOPEIA is bounded N. and W. by Cepheus, E. by Camelopardalus, and S. by Andromeda, and is one of the constellations of the Milky Way. It is readily distinguished by the figure of a *chair* inverted, of which two stars constitute the back, and four, in the form of a square, the body of the chair. It is on the opposite side of the pole from the Great Bear, and nearly at the same distance from it.

CAMELOPARDALUS (the GIRAFFE) is bounded N. by the Little Bear, E. by the head of the Great Bear, S. by Auriga and Perseus, and W. by Cassiopeia. Although this constellation occupies a large space, yet it has no conspicuous stars.

ANDROMEDA is bounded N. by Cassiopeia, E. by Perseus, S. by Pegasus, and W. by the Lizard. The direction of the figure is from S. W. to N. E., the head coming down within 30° of the equator, and being recognized by a star of the second magnitude, which forms the northeastern corner of the great square in Pegasus, to be described hereafter. At the distance of six or seven degrees from the head, are three conspicuous stars in a row, ranging from north to south, which lie in the *breast* of the figure; and about the same distance from these, and parallel to them, three more, which constitute the *girdle* of Andromeda. Near the northernmost of the three, is a faint, misty object, often mistaken for a comet, but is a nebula, and one of the most remarkable in the heavens.

PERSEUS is bounded N. by Cassiopeia, E. by Auriga, S. by Taurus, and W. by Andromeda. The figure extends from north to south, and is represented by a giant holding aloft a sword in his right hand, while his left grasps the *head of Medusa*,—a group of stars on the western side of the figure, embracing the celebrated star Algol. A series of bright stars descend along the shoulders and the waist, and there divide into the two legs. The western foot is 8° degrees north of the Pleiades. The eastern leg is bent at the knee, which is distinguished by a group of small stars. Near the sword handle, under Cassiopeia's chair, is a fine cluster of stars, so close together as scarce to be separable by the eye.

Auriga (the Wagoner) is bounded N. by Camelopardalus, E. by the Lynx, S. by Taurus, and W. by Perseus. He is represented as bearing on his left shoulder the little Goat *Capella*, a white and beautiful star of the first magnitude, (α Aurigæ), while Beta forms the right shoulder, 8° east of Capella. These two bright stars form, with the northern horn of the Bull, at the distance of 18°, an isosceles triangle.

Leo Minor (the Lesser Lion) is bounded N. by Ursa Major, E. by Coma Berenices, S. by Leo, and W. by the Lynx. It lies directly under the hind feet of the Great Bear, and over the sickle in Leo, and is easily distinguished. Four stars in the central part of the figure, from 4° to 5° apart, form a pretty regular parallelogram.

Canes Venatici (the Greyhounds). This constellation lies between the hind legs of the Great Bear on the west, and Bootes on the east; *Cor Caroli*, a solitary star of the third magnitude, 18° south of Alioth, in the tail of the Great Bear, will serve to mark this constellation.

Coma Berenices (Berenice's Hair) is a cluster of small stars, composing a rich group, 15° N. E. of Denebola, in the Lion's tail, in a line between this star and Cor Caroli, and half way between the two.

Bootes is bounded N. by Draco, E. by the Crown and the head of Serpentarius, S. by Virgo, and W. by Coma Berenices and the Hounds. It reaches for a great distance from north to south, the head being within 20° of the Dragon, and the feet reaching to the Zodiac. In the knee of Bootes is *Arcturus*, a star of the first magnitude. The next brightest star, *Beta*, is in the head of Bootes, 23° north of Arcturus, and 15° east of the last star in the tail of the Great Bear.

Corona Borealis (the Northern Crown) is bounded N. and E. by Hercules, S. by the head of Serpentarius, and W. by Bootes. It is formed of a semicircle of bright stars, six in number, of which *Gemma*, near the center of the curve, is of the second magnitude.

Hercules is bounded N. by Draco, E. by Lyra, S. by Ophiuchus, and W. by Corona Borealis. It is a very large constellation, and contains some brilliant objects for the telescope, although its components are generally very small. The figure lies north and south, with the head near the head of Ophiuchus, and the feet under the head of Draco. Being between the Crown and the Lyre, its locality is easily determined. The eastern foot of Hercules forms an isosceles triangle with the two southern stars of the trapezium in the head of Draco; while the head of Hercules is far in the south, within 15° of the equator, being 6° west of a similar star which constitutes the head of Ophiuchus.

Lyra (the Lyre) is bounded N. by the head of Draco, E. by the Swan, S. and W. by Hercules. Alpha Lyræ, or *Vega*, is of the first magnitude. It is accompanied by a small acute triangle of stars. Its color is a shining white, resembling Capella and the Eagle.

Cygnus (the Swan) extends along the Milky Way, below Cepheus, and immediately eastward of the Lyre, and has the figure of a large bird flying along the Milky Way from north to south, with outstretched wings and long neck. Commencing with the tail 25° east of Lyra, and following down the Milky Way, we pass along a line of conspicuous stars which form the body and neck of the figure; and then returning to the second of the series, we see two bright stars at eight or nine degrees on the right and left (the three together ranging across the Milky Way) which form the wings of the Swan. This constellation is among the few which exhibit some resemblance to the animals whose names they bear.

Vulpecula (the Little Fox) is a small constellation, in which a fox is represented as holding a goose in his mouth. It lies in the Milky Way, between the Swan on the north and the Dolphin and the Arrow on the south.

Aquila (the Eagle) stretches across the Milky Way, and is bounded N. by Sagitta, a small constellation which separates it from the Fox, E. by the Dolphin, S. by Antinous, and W. by Taurus Poniatowski (the Polish Bull), which separates it from Ophiuchus. It is distinguished by three bright stars in the neck, known as the "three stars," which lie in a straight line about 2° apart, on the eastern margin of the Milky Way. The central star is of the first magnitude. Its Arabic name is *Altair*.

Antinous lies across the equator, between the Eagle on the north and the head of Capricorn on the south.

DELPHINUS (the DOLPHIN) is situated east and north of Altair, and is composed of five stars of the third magnitude, of which four in the form of a rhombus, compose the head, and the fifth forms the tail.

PEGASUS (the FLYING HORSE) is a very large constellation, and is bounded N. by the Lizard and Andromeda, E. and S. by Pisces, W. by the Dolphin. The head is near the Dolphin, while the back rests on Pisces, and the feet extend toward Andromeda.

A large *square*, composed of four conspicuous members, one (*Markab*) of the first, and three others of the second magnitude, distinguish this constellation. The corners of the square are about 15° apart; the northeastern corner being in the head of Andromeda.

OPHIUCHUS is another very large constellation, the head being near the head of Hercules, and the feet reaching to Scorpio, the western foot being almost in contact with Antares. The figure is that of a giant holding a *serpent* in his hands. The head of the serpent is a little south of the Crown, and the tail reaches far eastward toward the Eagle.

416. Of the Constellations which lie south of the Zodiac, we shall notice only Cetus, Orion, Lepus, Monoceros, Canis Major, Canis Minor, Hydra, Crater, and Corvus.

CETUS (the WHALE) is distinguished rather for its extent than its brilliancy, occupying a large tract of the sky south of the constellations Pisces and Aries. The head is directly below the head of Aries, and the tail reaches westward 45°, being about 10° south of the vernal equinox. *Menkar* (*α Ceti*), the largest of its components, is situated in the mouth, 25° southeast of *α* Arietis; and *Mira* (*ο Ceti*) in the neck, 14° west of Menkar, is celebrated as a *variable* star, which exhibits different magnitudes at different times.

ORION is one of the most magnificent of the constellations, and one of those that have longest attracted the admiration of mankind, being alluded to in the book of Job, and mentioned by Homer. The head of Orion lies southeast of Taurus, 15° from Aldebaran, and is composed of a cluster of small stars. Two very bright stars, *Betalgeuse* of the first, and *Bellatrix* of the second magnitude, form the shoulders; three more, resembling the three stars of the eagle, compose the girdle; and three smaller stars, in a line inclined to the girdle, form the sword. *Rigel*, of the first magnitude, makes the west foot, but the corresponding star, 9° southeast of this, which is sometimes taken for the other foot, is above the knee, this foot being concealed behind the Hare. Orion's club is marked by three stars of the fifth magnitude, close together, in the Milky Way, just below the southern horn of the Bull. Orion is a favorite constellation with the practical astronomer, abounding, as it does, in addition to the splendor of its components, with fine nebulæ, double stars, and other objects of peculiar interest when viewed with the telescope. It embraces 70 stars, plainly visible to the naked eye, including two of the first, four of the second, and three of the third magnitude.

LEPUS (the HARE). Below Rigel, the western foot of Orion, is a small trapezium of stars, which forms the ears of the Hare; and an assemblage of nine stars, of the third and fourth magnitudes, south and east of these, make up the remaining parts of the figure.

CANIS MAJOR (the GREATER DOG) lies directly east of the Hare, and is highly distinguished by containing *Sirius*, the most splendid of all the fixed stars, which lies in the mouth of the figure. In the fore paw, 6° west of Sirius, is a star of the second magnitude (*β Canis Majoris*), and from 10° to 15° south of Sirius, is a collection of stars of the second and third magnitudes, which make up the hinder portions of the figure. The Egyptians, who anticipated the rising of the Nile by the appearance of Sirius in the morning sky, represented the constellation by the figure of a dog, the symbol of a faithful watchman.

CANIS MINOR (the LESSER DOG). About 25° north of Sirius, is the bright star *Procyon*, also of the first magnitude, which marks the side of the Lesser Dog. A star of the third magnitude (β), 4° northwest of this, in the head of the figure, forms with Procyon the lower side of an elongated parallelogram, of which Castor and Pollux, 25° north, form the upper side.

MONOCEROS is a large constellation, occupying the space between the Greater and the Lesser Dog, but has no conspicuous members.

HYDRA occupies a long space south of Leo, Virgo, and Libra. Its head, which is south of the fore paws of the Lion, consists of four stars of the fourth magnitude, of nearly uniform appearance; and about 15° S. E. of these is the *Heart* (*Cor Hydræ*), 23° south of Regulus. Resting on Hydra, and south of the hind feet of Leo, is *Crater* (the *Cup*), consisting of six stars of the fourth magnitude, arranged in the form of a semicircle; and a little further east, also perched on the back of Hydra, is *Corvus* (the *Crow*), the two brightest components of which are situated in one of the wings of the figure, in a line between Crater and Spica Virginis.

417. According to an intimation given in a note on p. 274, the constellations may be advantageously studied at four different periods of the year, as near the equinoxes and the solstices, according to the following directions. The latitude supposed is 41°.

Lesson I.—For the middle of *September*, from 8 to 10 o'clock. At 8 o'clock Scorpio is near setting in the S. W., Antares being 10° high. The bow of Sagittarius is seen on the eastern margin of the Milky Way, the arrow being directed to a point a little below Antares. At 9 o'clock, the horns of the Goat come upon the meridian; and at 10 o'clock, the western shoulder of Aquarius. The other shoulder, and the figure Y in the arm, may also be easily found from the description given on p. 277; also, the Pentagon, in Pisces, and Fomalhaut (the Southern Fish), a solitary bright star far in the south, only 16° above the horizon. The head of Aries appears in the east, and the Pleiades are but little above the horizon, while Aldebaran is just rising. Returning now to the west (at 10 o'clock), the Crown is seen a little north of west, about 20° high; Lyra is 30° west of the zenith; the Swan is nearly overhead: and following down the Milky Way, the Eagle is seen on its eastern margin over against Lyra on the western; and the Dolphin, a little eastward of the Eagle, and as far above the horns of Capricornus, as the latter are above the southern horizon. Following on east of the meridian, the great square in Pegasus may next be identified; and since the northeastern corner of the square is in the head of Andromeda, this constellation may next be learned; and then Perseus and Auriga, which appear still further east. Directly north of Perseus, is Cassiopeia's chair; and next to that we may take the Pole-star, the Little Bear, and the Great Bear, the *Dipper* only being traced for the present. Commencing now at the tail of the Dragon, we may trace round this figure between the two Bears to the head, which brings us back to Lyra and the head of Hercules. The boundaries of this constellation, and of Ophiuchus, which lies south of it, will end the first lesson.

LESSON II.—For the middle of *December*, from 7 to 10 o'clock. Of the constellations of the Zodiac, Taurus and Gemini are now favorably situated for observation in the east. At 7 o'clock, the tail of Cetus just reaches the meridian, its head being seen below the feet of Aries. Orion is just risen in the S. E. At 9 o'clock, just above the western horizon, are seen in succession from south to north, Aquarius, the Dolphin, the Eagle, the Lyre, and the Dragon's head. Between the Eagle and the Lyre, at a little higher altitude, we perceive the Swan, flying directly downward. Between the tail of the Swan and the Pole-star, is Cepheus; and from the pole, along the meridian, we trace Cassiopeia, the feet of Andromeda, the head of Aries, and the neck of the Whale. At 10 o'clock, Perseus has reached the meridian, the star Algol, in the head of Medusa, being directly overhead. The Pleiades are but little eastward of the zenith; and following along south from the pole, at the interval of from one to two hours east of the meridian, we may trace in succession, Camelopard, Auriga, Taurus, Orion, and the Hare. Turning along the eastern horizon, we find Canis Major, Monoceros, Canis Minor, the head of Hydra (just rising), Cancer, Leo, the sickle just appearing about 3° north of the east point. Leo Minor and Ursa Major complete the survey; and we may now advantageously trace out the various parts of the Great Bear, as described on p. 278; the two stars composing its hindmost paw being scarcely above the horizon.

LESSON III.—For the middle of *March*, from 8 to 10 o'clock. At 8 o'clock, we see the Twins nearly overhead, and Procyon and Sirius, at different intervals, toward the south. Along the west we recognize the neck and head of the whale, the head of Aries, and the head of Andromeda; next above these, Orion, Taurus, Perseus, Cassiopeia, and Cepheus; and north of the head of Orion, we see Auriga and Camelopard. In the S. W., Hydra is now fully displayed; and following on north, we obtain fine views of the Greater and the Lesser Lion, and the Great Bear. At 9 o'clock, Crater and Corvus appear in the S. E., on the back of Hydra; Virgo extends from Leo down to the horizon, Spica Virginis being about 5° high; and north of Virgo, we trace in succession Coma Berenices, Cor Caroli, Bootes, with Arcturus, and the Crown lying far in the N. E.

LESSON IV.—For the middle of *June*, from 9 to 10 o'clock. At 9 o'clock, Bootes, Corona Borealis, the head of Libra, the Serpent, and Scorpio, lie along on either side of the meridian. Castor and Pollux are just setting, and Leo is about an hour high. East of Leo, Virgo is seen extending along toward the meridian, Spica being about 30° above the southern horizon. North of Leo and Virgo we recognize Leo Minor, Coma Berenices, Cor Caroli, and Ursa Major. At 10 o'clock, we trace along the eastern side of the meridian, Draco, Hercules, and Ophiuchus; and east of these, the Lyre, the Eagle, Antinous, Sagittarius, and Capricornus. North of the Eagle, and round to the east, we find Cepheus and Cassiopeia, Andromeda rising in the northeast, Pegasus in the east, and Aquarius in the southeast. Thus we may advantageously complete a review of the constellations.

CHAPTER II.

DOUBLE STARS—TEMPORARY STARS—VARIABLE STARS—CLUSTERS AND NEBULÆ.

418. THE view hitherto taken of the starry heavens presents little that is new, since most of the Constellations, visible in our latitude, and the most conspicuous of the individual stars, have been known from antiquity. But the objects to be described in the present chapter, are chiefly such as have been discovered by modern astronomy, aided by the powerful telescopes which, since the time of Sir William Herschel, have been directed to the heavens. Different orders and systems of stars have been brought to light, and a new and still more wonderful class of bodies, called Nebulæ, have been reached in the depths of the stellar universe.

419. The introduction into practical astronomy of Herschel's great Forty-feet Reflector, in 1789, was a great event in the study of the stars. This instrument in its previous humble forms had been very little employed upon the stars, they being supposed to be too remote for its powers, which seemed only suited to nearer worlds, as the sun and planets. It was not, however, an increase of magnifying power that was wanted for researches on these distant objects, but an increase of *light*, by which a few scattered rays sent to us from bodies hidden in the depths of space, might be collected in such numbers, and directed into the eye, as would render visible objects otherwise invisible, not because they do not transmit us *any* light, but because not enough of what they transmit enters the small pupil of the eye for the purposes of distinct vision. Telescopes of great aperture, therefore, by collecting a large beam of light and conveying it to the eye, greatly enlarge the powers of this organ, and enable it to penetrate porportionally further into the most distant regions of the universe. Sir W. Herschel himself made wonderful progress in the knowledge of the starry heavens, and by his own researches discovered a large portion of those bodies which we are now to describe; and his son, Sir John Herschel, has cultivated, with great success, the same field, and especially by a residence of five years at the Cape of Good Hope, devoted assiduously to observations with large instruments, has greatly augmented our knowledge of the stellar systems of the southern hemisphere. Moreover, telescopes of still greater power than that of the elder Herschel, and especially instruments capable of nicer angular measurements, have recently enriched the department of practical astronomy. The most remarkable of these are the grand *Reflector* constructed by Lord Rosse, an Irish nobleman, and the great *Refractors* belonging respectively to the Pulkova and Cambridge Observatories. Lord Rosse's telescope considerably exceeds in dimensions and in power the forty-feet reflector of Sir W. Herschel, being 50 feet in focal length, and having a diameter of 6 feet, whereas that of the Herschelian telescope was only 4 feet. This unexampled magnitude makes this instrument superior to all others in *light*, and fits it pre-eminently for observations on the most remote and obscure celestial objects, such as the faintest nebulæ. But its unwieldy size, and its liability to loss of power, by the tarnishing or temporary blurring of the great speculum, will render it far less available for actual research than the great refractors which come into competition with it. Until recently, it was thought impossible to form perfect achromatic object-glasses of more than about five inches diameter; but they have been successively enlarged, until we can no longer set bounds to the dimensions which they may finally assume. The Pulkova telescope (at St. Petersburg) has a clear aperture of about 15 inches, and a focal length of 22 feet. The telescope recently acquired by Harvard University, is perhaps the finest refractor hitherto constructed. It was made by the same artists, and upon the same scale with that, but its performances are thought even to exceed those of the Pulkova instrument. We now proceed to review some of the discoveries among the stars, which the researches made with such instruments as the foregoing have brought to light.

DOUBLE STARS.

420. Double Stars are those which appear single to the naked eye, but are resolved into two by the telescope; or if not visible to the naked eye, are seen in the telescope very close together. Sometimes three or more stars are found in this near connection, constituting *triple* or *multiple* stars.* Castor, for example, when seen by the naked eye, appears as a single star; but in a telescope, even of moderate powers, it is resolved into two stars, between the third and fourth magnitudes, within 5″ of each other. These two stars are of nearly equal size, but frequently one is exceedingly small in comparison with the other, resembling a satellite near its primary, although in distance, in light, and in other characteristics, each has all the attributes of a star, and the combination, therefore, cannot be that of a planet with a satellite. The distance between these

* See several figures of double and multiple stars, in Plate III. at the end of the volume.

objects varies from a fraction of a second to thirty-two seconds. In some cases, the extreme closeness, and the exceeding minuteness of double stars, require, for their separation, the best telescope, united with the most acute powers of observation. Indeed, certain of these objects are regarded as the severest *tests* both of the excellence of the instrument, and of the skill of the observer.

421. When Sir William Herschel began his observations on double stars, about the year 1780, he was acquainted with only 4. By his own researches he extended the number to 2,400. Sir John Herschel, Sir James South, and M. Struve, the great Russian astronomer, prosecuted the same line of research; and when Sir John Herschel left England for the Cape of Good Hope, in 1833, the whole number of double stars enrolled was 3,346; and this number was increased, by that eminent astronomer, by adding those of the southern hemisphere, to 5,542. It appears, therefore, that the number of double stars considerably exceeds all the stars visible to the naked eye. In some instances, this proximity arises undoubtedly from the two members lying nearly in the same line of vision, and therefore being projected very near to each other on the face of the sky; but in most cases the double stars are proved to have a physical relation to each other, and are therefore said to be *physically* double, while the former are said to be *optically* double. There is no longer any doubt that among the stars are separate *systems*, in which two, three, and even in one instance at least, six stars are bound together in relations of mutual dependence, suns with suns, as the members of the solar system compose an individual province in the great empire of nature. A star in Orion's sword (*Theta Orionis*) has been for some time known as a *quadruple* star, the members of which form a small trapezium; and recent observations have detected in two of these, severally, companions of extreme minuteness, the whole composing a figure like the following:

Many of the double stars are distinguished by the components exhibiting different colors, often finely contrasted with each other; as orange with blue or green, yellow with blue, and white with purple. Gamma Andromedæ is a close double star, the components of which are both green. Insulated stars of a red color, almost as deep as that of blood, occur in many parts of the heavens, but no green or blue star of any decided hue has ever been noticed unassociated with a companion brighter than itself.*

422. TEMPORARY STARS.

TEMPORARY STARS are new stars which have suddenly made their appearance, and after a certain interval, as suddenly disappeared, and returned no more. It was the appearance of a new star of this kind, 125 years before the Christian era, that prompted Hipparchus to form a catalogue of the stars, the first on record. Such also was the star which suddenly shone out, A. D. 389, in the Eagle, as bright as Venus, and after remaining three weeks, disappeared entirely. At other periods, at distant intervals, similar phenomena have presented themselves. Thus the appearance of a new star in 1572 was so sudden, that Tycho Brahe, returning home one evening, was surprised to find a collection of country people gazing at a star which he was sure did not exist half an hour before. It was then as bright as Sirius, and continued to increase until it surpassed Jupiter when brightest, and was visible at mid-day. In a month it began to diminish, and in three months afterward it had entirely disappeared. Some stars are now missing which were registered in the older catalogues. In one instance, at least (that of Neptune), the supposed star has proved to have been a planet.

423. VARIABLE STARS.

VARIABLE STARS are those which undergo a periodical change of brightness. One of the most remarkable is the star *Mira*, in the neck of the Whale (*Omicron Ceti*). It appears once in 11 months, remains at its greatest brightness about a fortnight, being then, on some occasions, equal to a star of the second

* Herschel.

magnitude. It then decreases about three months, until it becomes completely invisible, and remains so about five months, when it again becomes visible, and continues increasing during the remaining three months of its period.

Another very remarkable variable star is *Algol* (β Persei). It is suddenly visible as a star of the second magnitude, and continues such for 2d. 14h., when it begins rapidly to diminish in splendor, and in about 3½ hours is reduced to the fourth magnitude. It then begins again to increase, and in 3½ hours more, is restored to its usual brightness, going through all its changes in less than three days. This remarkable law of variation appears strongly to suggest the revolution round it of some opaque body, which, when interposed between us and Algol, cuts off a large portion of its light. It is (says Sir J. Herschel) an indication of a high degree of activity in regions where, but for such evidence, we might conclude all to be lifeless. Our sun requires almost nine times this period to perform a revolution on its axis. On the other hand, the periodic time of an opaque revolving body, sufficiently large, which would produce a similar temporary obscuration of the sun, seen from a fixed star, would be less than fourteen hours.

The duration of these periods is extremely various. While that of β Persei, above mentioned, is less than three days, others are more than a year, and others many years.

124. CLUSTERS AND NEBULÆ.

In various parts of the firmament are seen large groups or CLUSTERS, which, either by the naked eye, or by the aid of the smallest telescope, are perceived to consist of a great number of small stars. Such are the Pleiades, Coma Berenices, and Præsepe, or the Beehive in Cancer. The Pleiades, or *Seven Stars*, as they are called, in the neck of Taurus, is the most conspicuous cluster. When we look *directly* at this group, we can not distinguish more than six stars, but by turning the eyes a little to one side,* we discover that there are many more.

* Indirect vision is far more delicate than direct. Thus we can see the Zodiacal Light or a Comet's tail much more distinctly and of greater length, if, instead of looking directly *at* it, we turn the eyes to various points *near* it, the *attention* all the while being given to the object itself.

The telescope only can, however, display the real magnificence of the Pleiades. (See Plate III., Fig. 1.) Coma Berenices has fewer stars, but they are of a larger class than those which compose the Pleiades. The Beehive, or Nebula of Cancer, is one of the finest objects of this kind for a small telescope, being, by its aid, converted into a rich congeries of shining points. A cluster in the sword-handle of Perseus, below Cassiopeia's chair, though but a dim speck to the naked eye, is a very elegant object to a large telescope, being separated into bright and beautiful stars, embracing several distinct subordinate clusters of exceedingly minute stellar points. The head of Orion affords an example of another cluster, though less remarkable than the others.

425. Nebulæ are faint misty objects seen in various parts of the firmament, always maintaining a fixed position, which resemble comets, or patches of fog. The Galaxy, or Milky Way, presents a constant succession of large nebulæ. Of the individual nebulæ, seen by the naked eye, the most conspicuous is that near the girdle of Andromeda. It is the oldest known nebula, having attracted the attention of star-gazers as early as the beginning of the tenth century,* although it is commonly said to have been discovered by Simon Marius, in 1612. No powers of the telescope have been able to resolve this into separate stars, although the great Cambridge telescope reveals a vast number of stars, more than 1,500, of various degrees of brightness, scattered over its surface; but these appear not to belong to the nebula itself, which has hitherto afforded no evidence of resolution.† Its dimensions are astonishingly great, since it covers a space of a quarter of a degree in diameter; and we must bear in mind that, at such a distance as the fixed stars, a space of 15′ implies an immense extent. Its figure is oval, and *elliptical* nebulæ constitute a common variety among the figures which these bodies exhibit. (See Plate III., Fig. 2, for a representation of the great nebula of Andromeda.) Another very common figure are the *globular* nebulæ. A grand specimen of this variety may be easily

* Smyth's Cycle, ii., p. 15.
† Memoirs of the Amer. Acad., vol. iii.

found in the constellation Hercules, between Zeta and Eta. Draw a line from Lyra to Gemma of the Crown, and 3° above the center of that line will be the place of this nebula. When viewed with a small telescope, it exhibits only a circular cloud (Plate III., Fig. 3, *a*), but to a more powerful instrument it reveals its real glories in a form truly exciting to the beholder (Fig. 3, *b*). About 4000 nebulæ have been detected and described, of which about 1,700 have recently been added by Sir John Herschel, from his Results of Observations at the Cape of Good Hope Among the latter are two remarkable spots, well known to navigators, situated near the south pole, called *Magellanic clouds* by sailors, but by astronomers, the *Nubecula Major* and the *Nubecula Minor*. They are found to consist of a wonderful collection of nebulæ, the greater embracing 278 nebulæ, and the lesser 37. Both together compose a most magnificent assemblage. In the *sword of Orion* is a celebrated nebula, long known, which, until recently, had resisted all attempts to resolve it into stars; but the great Reflector of Lord Rosse, and more recently the great Refractor of the Cambridge Observatory, have succeeded in a partial resolution, at least, of this grand object, and have authorized the anticipation that, with a small increase of telescopic power, the whole will be shown to consist of an immense collection of exceedingly minute stars.

These great telescopes, by the superior light they afford, display their peculiar powers in this department of astronomy, and those astronomers who, for the first time, have gazed at these sidereal pictures as seen in the "Leviathan" of Lord Rosse, have expressed, in glowing terms, their mingled delight and astonishment. The perfect forms, and strange but symmetrical configurations, exhibited by these instruments, of nebulæ that were before seen of irregular or fantastic shapes, afford grounds for believing that such irregularities are often if not always owing to the objects being but partly developed. Thus the *Crab Nebula* of Lord Rosse (Plate III., Fig. 4) had been long known as a faint, ill-defined nebula of an elliptical shape; but the higher powers of that instrument exhibit the before-concealed appendages which are essential to the completeness of the figure. The *Whirlpool Nebula* of Rosse (Plate III., Fig. 5), when seen in separate parts, exhibited no signs of order or symmetry; but when viewed with the great Reflector, it develops the wonderful structure of a perfect spiral.

426. Nebulæ were formerly divided into two classes, *resolvable* and *irresolvable*, the former term implying that the body was shown by the telescope to consist of stars, and the latter implying that the body is not composed of stars, but of a shining, cloudy kind of matter diffused throughout the mass. Astronomers, at present, include all resolvable nebulæ under the head of *clusters*, appropriating the term nebulæ exclusively to such of these bodies as have never been resolved. The question whether this distinction is not merely relative to the powers of the telescope, and whether, on the increase of these powers, this class of bodies would not all be resolved into stars, is not easily determined, since the same increase of telescopic power which converts existing nebulæ into clusters, brings to light a greater number of those which are irresolvable.

These remote objects of the universe occasionally exhibit traces of that regard to beauty which everywhere, in these nether worlds, characterizes the works of the Creator. In the *Cross*, a brilliant constellation of the southern hemisphere, for example, is a cluster surrounding the star *Kappa Crucis*, which consists of about 110 stars from the seventh magnitude downward, eight of the more conspicuous of which are colored with various shades of red, green, and blue, so as to give to the whole the appearance of a rich piece of jewelry.

427. *Nebulous stars* are such as exhibit a sharp and brilliant star, surrounded by a disk or atmosphere of nebulous matter. These atmospheres in some cases, present a circular, in others an oval figure; and in certain instances, the nebula consists of a long, narrow, spindle-shaped ray, tapering away at both ends to points. *Annular Nebulæ* (Ring-shaped) are among the rarest objects in the heavens. The most conspicuous of this class is in the constellation Lyra, between the stars Beta and Gamma, about 6° S. E. of Alpha Lyræ. This remarkable object is believed to be in fact a resolvable nebula or cluster, and yet the greatest powers of the telescope hitherto applied have only effected such changes as are regarded as giving *signs* of resolvability, but its perfect resolution has not been attained. Should it be achieved by an increased power of the instrument, astronomers look for a splendid coronet of stars, more glorious, perhaps, than any thing hitherto discovered in the starry heavens.

Planetary Nebulæ constitute another variety, and are very remarkable objects. They have, as their name imports, exactly the appearance of planets. Whatever may be their nature, they must be of enormous magnitude. One of them is to be found in the parallel of ν Aquarii, and about 5m. preceding that star. Its apparent diameter is about 20″. Another in the constellation Andromeda, presents a visible disk of 12″, perfectly defined and round. Granting these objects to be equally distant from us with the stars, their real dimensions must be such as, on the lowest computation, would fill the orbit of Uranus. It is no less evident that, if they be solid bodies, of a solar nature, the intrinsic splendor of their surfaces must be almost infinitely inferior to that of the sun. A circular portion of the sun's disk, subtending an angle of 20″, would give a light equal to 100 full moons; while the objects in question are hardly, if at all, discernible with the naked eye.*

428. The *Milky Way*, or *Galaxy*, is a well-known luminous zone, encircling the sphere nearly in the direction of a great circle. Near the Swan, in the northern sky, it is seen to be divided into two bands, which remain asunder for 150°, and then reunite. The Galaxy owes its peculiar appearance to the blended light of myriads of small stars too minute to be individually recognized by the naked eye, but which are seen in their true character by a telescope of only moderate powers. Sir William Herschel estimated, that, on one occasion, in forty-one minutes, no less than 258,000 stars passed through the small field of his telescope.† In approaching the border of the Milky Way, there is found a regular but rapid increase in the number of stars, even before entering the limits of the luminous zone itself. Sir J. Herschel computes the whole num-

* Herschel.

† Plate II., Fig. 1, exhibits a telescopic view of a part of the southern portion of the Milky Way.

ber of stars in the Milky Way at *five and a half millions*, including such only as are visible in his twenty-feet reflector. The Galaxy is supposed to be one of the numerous nebulæ, and the sun of our own solar system to be one of the stars which compose it. It appears comparatively large to us, and extends entirely round the heavens, only because we are in the midst of it, and see it projected in different directions from us.

CHAPTER III.

MOTIONS OF THE FIXED STARS—DISTANCES—NATURE.

429. In 1803, Sir William Herschel first determined and announced to the world, that there exist among the stars separate systems, composed of two stars, revolving about each other in regular orbits. These he denominated *Binary Stars*, to distinguish them from other double stars where no such motion is detected, and whose proximity to each other may possibly arise from casual juxtaposition, or from one being in the range of the other. At present, more than a hundred of the binary stars are known, and as the number of such revolutions known among the double stars is constantly increasing as the times of comparison increase, it may be anticipated that, in after ages, so large a proportion of all the double stars will be found to possess this character, as to authorize the belief that they universally consist of subordinate systems, of which the members have a revolution around a common center of gravity. The *periodic times* of the binary stars are very various. While some (as ζ *Herculis*, and η *Coronæ*) complete their revolutions in 30 or 40 years, others (as γ *Virginis*) require more than 170, and others still (as 65 *Piscium*) take up the long period of 3000 years.* Their orbits are in general more eccentric than those of the planets. That of Gamma Virginis, including the relative positions of the two components

* Smyth's Cycle, i., p. 300.

from 1837 to 1860, is figured on Plate II. as drawn by Mr. E P. Mason, in 1840.*

430. The revolutions of the binary stars have assured us of this most interesting fact, that *the law of gravitation extends to the fixed stars.* Before these discoveries, we could not decide, except by a feeble analogy, that this law transcended the bounds of the solar system. Indeed, our belief of the fact rested more upon our idea of unity of design in all the works of the Creator, than upon any certain proof; but the revolution of one star around another, in obedience to forces which must be similar to those that govern the solar system, establishes the grand conclusion, that the law of gravitation is truly the law of the material universe.

We have the same evidence (says Sir John Herschel) of the revolutions of the binary stars about each other, that we have of those of Saturn and Uranus about the sun; and the correspondence between their calculated and observed places in such elongated ellipses, must be admitted to carry with it a proof of the prevalence of the Newtonian law of gravity in their systems, of the very same nature and cogency as that of the calculated and observed places of comets round the center of our own system.

But (he adds) it is not with the revolutions of bodies of a planetary or cometary nature round a solar center that we are now concerned; it is with that of sun around sun, each, perhaps, accompanied with its train of planets and their satellites, closely shrouded from our view by the splendor of their respective suns, and crowded into a space, bearing hardly a greater proportion to the enormous interval which separates them, than the distances of the satellites of our planets from their primaries, bear to their distances from the sun itself.

431. *Some of the fixed stars appear to have a* Proper Motion, *or a real motion in space.*

* Sir John Herschel had computed the orbit of γ Virginis, and had given it at 625 years. Mason, from a discussion of all the observations, published to the date of 1838, combined with his own of 1840, found that this period was too great, and assigned as the true period 171 years, which is now acknowledged by the highest authorities, and even by Herschel himself, to be nearly its real time of revolution.

The *apparent* change of place in the stars arising from the precession of the equinoxes, the nutation of the earth's axis, the diminution of the obliquity of the ecliptic, and the aberration of light, have been already mentioned; but after all these corrections are made, changes of place still occur, which cannot result from any changes in the earth, but must arise from changes in the stars themselves. Such motions are called the *proper motions* of the stars. Nearly 2000 years ago, Hipparchus and Ptolemy made the most accurate determinations in their power of the relative situations of the stars, and their observations have been transmitted to us in Ptolemy's Almagest; from which it appears that the stars retain at least very *nearly* the same places now as they did at that period. Still, the more accurate methods of modern astronomers, have brought to light minute changes in the places of certain stars which force upon us the conclusion, *either that our solar system causes an apparent displacement of certain stars, by a motion of its own in space, or that they have themselves a proper motion.* Possibly, indeed, both these causes may operate.

432. If the sun, and of course the earth which accompanies him, is actually in motion, the fact may become manifest from the apparent approach of the stars in the region which he is leaving, and the recession of those which lie in the part of the heavens toward which he is traveling. Were two groves of trees situated on a plain at some distance apart, and we should go from one to the other, the trees before us would gradually appear further and further asunder, while those we left behind would appear to approach each other. Some years since, Sir William Herschel supposed he had detected changes of this kind among two sets of stars in opposite points of the heavens, and announced that the solar system was in motion toward a point in the constellation Hercules.* As, for many years after this announcement, other astronomers failed to find evidence of such a motion of the solar system, the doctrine was generally discredited, until, within a few years, new and very refined researches have been instituted by several of the most eminent astronomers, which have fully confirmed the observations of

* Phil. Trans., 1783, 1805, and 1806.

Herschel. The great Russian astronomer, Struve, by a comparison of the best observations, finds the exact point toward which the solar system is moving is in a line which joins the two stars π and μ Herculis,*—a point which can be easily found on the celestial globe, and thence transferred to the heavens. (Right ascension 259°, declination $34\frac{1}{2}$°.) The researches of the younger Struve have conducted him to the *velocity* with which the solar system is moving in space. For having found that the arc traversed by the sun in a year is 0″.3392, if viewed at the mean distance of the stars of the first magnitude, and having previously ascertained that the mean parallax of the stars of this class amounts to 0″.209, he infers that the space through which the sun moves annually is 154,000,000 miles. Great as this space is, yet it may be remarked that it is only about one-fourth that traversed by the earth in its revolution around the sun. Within the comparatively short period during which these observations on the solar motion have been continued, the direction appears rectilinear; but all analogy leads to the belief that it is in fact a motion of revolution, although on account of the immense size of the orbit, and, consequently, its small curvature, many years will be requisite in order to determine the deviation from the line of the tangent.†

433. When we reflect on the immense distance of the stars, we may readily believe that they may be in fact in rapid motion, and yet appear quiescent; as a distant ship, under full sail, appears at rest, although actually moving at the rate of ten knots an hour. Thus we have seen above that a motion of the sun in space, as seen from the nearest fixed stars, would make it describe an arc of only about one-third of a second annually, although traversing a space of 154 millions of miles. But a small change in the place of a star in a single year may, in a long series of years, accumulate to a very sensible amount. For example, the latitudes of the three bright stars, Sirius, Arcturus, and Aldebaran, were determined by Hipparchus 130 years before the Christian era, and their assigned places are transmitted to us in the Almagest of Ptolemy. About the year 1700, Dr. Halley found that these stars had, during the

* Études d'Astron. Stellaire, p. 108. † Grant's Hist. Phys. Ast., p. 557.

interval of nearly 2000 years, moved southerly through the spaces respectively of 37′, 42′, and 33′. The immense pains that have of late years been bestowed upon catalogues of the stars, and especially of particular portions of the heavens, with the view of furnishing to after ages the most accurate data for comparison, will enable future astronomers to study the proper motions of the stars with far greater advantages than the present generation enjoys. In most cases where a proper motion in certain stars has been suspected, its annual amount has been so small, that many years are required to assure us that the effect is not owing to some other than a real progressive motion in the stars themselves; but in a few instances the fact is too obvious to admit of any doubt. A small star in the leg of the Great Bear has an annual motion away from the neighboring stars of 7″; and the two stars 61 Cygni, which are nearly equal, have remained constantly at the same, or nearly at the same distance of 15″ for at least fifty years past. Meanwhile they have shifted their local situation in the heavens 4′ 23″, the annual proper motion of each star being 5″.3, by which quantity this system is every year carried along in some unknown path, by a motion which for many centuries must be regarded as uniform and rectilinear. A greater proportion of the double stars than of any other indicate proper motions, especially the binary stars, or those which have a revolution around each other. Among stars not double, and no way differing from the rest in any other obvious particular, μ Cassiopeiæ has a proper motion, amounting to nearly 4″ annually; and another obscure star has been recently found to have a motion of nearly 8″.*

434. DISTANCES OF THE FIXED STARS.

It has long been considered one of the highest problems that can be proposed to the human mind, to measure the distance to any of the fixed stars. Nothing more would be necessary than to measure the horizontal parallax, if it were possible; but this is now known to be less than the 70,000th part of a second for the nearest star, and therefore utterly inappreciable by any method of measurement. For measuring the distances

* Herschel's Outlines (Ed. 1851).

of the sun and planets, the *diameter of the earth* furnishes the base line (Art. 87). The length of this line being known, and likewise the horizontal parallax of the body whose distance is sought, we readily obtain the distance by the solution of a right-angled triangle (Art. 80, Fig. 6). But any star viewed from the opposite sides of the earth would appear from both stations to occupy precisely the same situation in the celestial sphere, and of course it would exhibit no horizontal parallax. But astronomers have endeavored to find a parallax in some of the fixed stars, by taking the *diameter of the earth's orbit* as a base line. Yet even a change of position, amounting to 190 millions of miles, has, until within a few years, proved insufficient to alter the apparent place of a single fixed star, from which it was concluded that the fixed stars have not even any *annual* parallax; or that the angle subtended by the semi-diameter of the earth's orbit, at the nearest fixed star, is insensible. The errors to which instrumental measurements are subject, arising from defects of the instruments themselves, from errors of refraction, of aberration, of precession, of nutation, and from imperfections of observation, are such, that the angular determinations of celestial arcs, it was supposed, could not be relied on to less than 1″; and the change of place in any star that had been examined for parallax being less than one second when viewed at opposite extremities of the earth's orbit, the conclusion was, that the parallax of the fixed stars, if any exist, is too minute ever to be measured by instruments. According to this, the diameter of the earth's orbit, when viewed from the nearest fixed star, would be insensible; the spider-line of the telescope would more than cover it.

Fig. 80.

c

a b

Taking, however, the annual parallax at 1″, let *ab* (Fig. 80) represent the radius of the earth's orbit, and *c* a fixed star, the angle at *c* being 1″, and the angle at *b* a right angle; then,

Sin 1″ : Rad :: 1 : 200,000, nearly.

Hence the hypotenuse of a triangle whose vertical angle is 1″, is about 200,000 times the base; consequently, the distance in question *must exceed* 95,000,000 × 200,000 = 190,000,000 × 100,000, or one hundred thousand times one hundred and ninety millions of miles. Of a distance so vast we can form no adequate conceptions, and attempt to measure it only by the time that light (which moves more than 192,000 miles per second) would take to traverse it. Now,

192,000 : 1*s.* :: 19,000,000,000,000 : 3.1 years.

435. After many fruitless and delusory efforts to measure the immense interval that separates us from the fixed stars, the great Prussian astronomer, Bessel, in the year 1838, determined this interesting and important element, by observations on a double star in the Swan (61 Cygni). This star was selected for the following reasons: first, it was known to have a great *proper motion* (Art. 433), indicating a comparatively great proximity to our system; secondly, situated as it is among the circumpolar stars, observations could be made upon it nearly every night in the year; and thirdly, the great number of small stars in the immediate neighborhood, furnished the opportunity of selecting favorable stationary points from which (inasmuch as these more remote objects might be considered as entirely devoid of parallax) any changes of place in the nearer, in consequence of an annual parallax, might be readily estimated. By observations of the last degree of refinement, conducted for a period of several years, a parallax was decisively indicated, amounting to about one-third of a second; or, more exactly, to 0″.3483, implying a distance of 592,200 times the mean distance of the earth from the sun, or a space which it would take light, moving at the rate of twelve millions of miles per minute, *nine and a quarter years to traverse.* Perhaps the best way to conceive of this distance, is to compare it with the dimensions of the solar system, as represented by the diagram (note, p. 178). The distance from the sun to Neptune being 30 feet, 61 Cygni should be placed 110 *miles* off. Thus isolated are the systems of the universe from each other.

The observations of Bessel enabled him to estimate also the period of revolution of the two stars composing the binary system of 61 Cygni, and the dimensions of the orbit, and he found the periodic time about 540 years, and the length of the orbit about two and a half times that of Uranus. Knowing the distance and proper motion of the star, we can now obtain its *velocity*, so far as it is perpendicular to our line of vision. This is found to be about forty-four miles per second—more than double that of the earth in its orbit—amounting to about one thousand millions of miles per annum.

On account of the smallness of the supposed parallax thus found, it would not be unreasonable still to entertain a lingering suspicion, that it is nothing more than the unavoidable imperfection of instrumental measurements, as proved to be the case in previous attempts to find the same element; but the most satisfactory evidence which the world can have that such is not the fact in the present instance, but that the parallax is truly found, is that the most celebrated astronomers of the age, after rigorous scrutiny, have acknowledged the reality and soundness of the determination. Our confidence that the parallax of 61 Cygni was truly determined by Bessel, is strengthened by the fact that a separate determination recently made by Peters at the Pulkova Observatory, gives almost precisely the same result, that of Bessel being 0''.348, and that of Peters 0''.349. In the case of several stars still more distant, the parallax has been found, with more or less probability, but with sufficient to command the general confidence of astronomers. Thus, the parallax of Arcturus, Alpha Lyræ, and Polaris, were also found by Peters to be respectively 0''.127, 0''.123, 0''.067, that of the Pole-star being only one-fifth as great as that of 61 Cygni; and, consequently, if light would require 9¼ years to come from that star, it would require more than 46 years to come to us from the Pole-star. A star in the southern hemisphere (*α Centauri*) indicates a parallax of about 1'', and hence appears at present the nearest of the fixed stars.

436. NATURE OF THE STARS.

The stars are bodies greater than our earth. If this were not the case they could not be visible at such an immense distance. Dr. Wollaston, a distinguished English philosopher, attempted to estimate the magnitudes of certain of the fixed stars from the light which they afford. By means of an accurate photometer (an instrument for measuring the relative intensities of light) he compared the light of Sirius with that of the sun. He next inquired how far the sun must be removed from us in order to appear no brighter than Sirius. He found the distance to be 141,400 times its present distance. But Sirius is more than 200,000 times as far off as the sun (Art. 434). Hence he inferred that, upon the lowest computation, Sirius must actually give out twice as much light as the sun; or that, in point of splendor, Sirius must be at least equal to two suns. Indeed, he has rendered it probable that the light of Sirius is equal to fourteen suns.

437. *The fixed stars are suns.* We have already seen that they are large bodies; that they are immensely further off than the furthest planet; that they shine by their own light, as is evident by the nature of the light as tested by polarization; in short, that their appearance is, in all respects, the same as the sun would exhibit if removed to the region of the stars. Hence we infer that they are bodies of the same kind with the sun.

438. We are justified therefore by a sound analogy, in concluding that the stars were made for the same end as the sun, namely, as the centers of attraction to other planetary worlds, to which they severally dispense light and heat. Although the starry heavens present, in a clear night, a spectacle of ineffable grandeur and beauty, yet it must be admitted that the chief purpose of the stars could not have been to adorn the night, since by far the greater part of them are wholly invisible to the naked eye; nor as landmarks to the navigator, for only a very small proportion of them are adapted for this purpose; nor, finally, to influence the earth by their attractions, since their distance renders such an effect entirely insensible. If they are suns, and if they exert no important agencies upon our world, but are bodies evidently adapted to the same purpose as our sun, then it is as rational to suppose that they were made to give light and heat, as that the eye was made for seeing and the ear for hearing. It is obvious to inquire next, to what they dispense these gifts, if not to planetary worlds; and why to planetary worlds, if not for the use of percipient beings? We are thus led, almost inevitably, to the idea of a *Plurality of Worlds;* and the conclusion is forced upon us, that the spot which the Creator has assigned to us is but an humble province of his boundless empire.*

* See this argument, in its full extent, in *Dick's Celestial Scenery.*

CHAPTER IV.

OF THE SYSTEM OF THE WORLD.

439. *The arrangement of all the bodies that compose the material universe, and their relations to each other, constitutes the System of the World.*

It is otherwise called the Mechanism of the Heavens; and, indeed, in the System of the World, we figure to ourselves a machine, all the parts of which have a mutual dependence, and conspire to one great end. "The machines that are first invented (says Adam Smith) to perform any particular movement, are always the most complex; and succeeding artists generally discover that with fewer wheels and with fewer principles of motion than had originally been employed, the same effects may be more easily produced. The first systems, in the same manner, are always the most complex; and a particular connecting chain or principle is generally thought necessary to unite every two seemingly disjointed appearances; but it often happens that *one great connecting principle* is afterward found to be sufficient to bind together all the discordant phenomena that occur in a whole species of things." This remark is strikingly applicable to the origin and progress of systems of astronomy.

440. From the visionary notions which are generally understood to have been entertained on this subject by the ancients, we are apt to imagine that they knew less than they actually did of the truths of astronomy. But Pythagoras, who lived 500 years before the Christian era, was acquainted with many important facts in our science, and entertained many opinions respecting the System of the World which are now held to be true. Among other things well known to Pythagoras were the following:

1. The principal *constellations.* These had begun to be formed in the earliest ages of the world. Several of them bearing the same names as at present are mentioned in the writings of Hesiod and Homer; and the "sweet influences of

the Pleiades" and the "bands of Orion," are beautifully alluded to in the Book of Job.

2. *Eclipses.* Pythagoras knew both the causes of eclipses and how to predict them;* not indeed in the accurate manner now employed, but by means of the Saros (Art. 233).

3. Pythagoras had divined the true system of the world, holding that the sun, and not the earth (as was generally held by the ancients, even for many years after Pythagoras), is the center around which all the planets revolve, and that the stars are so many suns, each the center of a system like our own.† Among lesser things, he knew that the earth is round; that its surface is naturally divided into five zones; and that the ecliptic is inclined to the equator. He also held that the earth revolves daily on its axis, and yearly around the sun; that the galaxy is an assemblage of small stars; and that it is the same luminary, namely, Venus, that constitutes both the morning and the evening star, whereas all the ancients before him had supposed that each was a separate planet, and accordingly the morning star was called Lucifer, and the evening star Hesperus.‡ He held also that the planets were inhabited, and even went so far as to calculate the size of some of the animals in the moon.§ Pythagoras was so great an enthusiast in music, that he not only assigned to it a conspicuous place in his system of education, but even supposed the heavenly bodies themselves to be arranged at distances corresponding to the diatonic scale, and imagined them to pursue their sublime march to notes created by their own harmonious movements, called the "music of the spheres;" but he maintained that this celestial concert, though loud and grand, is not audible to the feeble organs of man, but only to the gods.

441. With few exceptions, however, the opinions of Pythagoras on the System of the World, were founded in truth. Yet they were rejected by Aristotle and by most succeeding astronomers down to the time of Copernicus, and in their place was substituted the doctrine of *Crystalline Spheres*, first taught by

* Long's Astronomy, ii., p. 671.
† Library of Useful Knowledge, *History of Astronomy.*
‡ Long's Ast., ii., p. 673. § Edin. Encyclopædia.

Eudoxus. According to this system, the heavenly bodies are set like gems in hollow solid orbs, composed of crystal so pellucid that no anterior orb obstructs in the least the view of any of the orbs that lie behind it. The sun and the planets have each its separate orb; but the fixed stars are all set in the same grand orb; and beyond this is another still, the *Primum Mobile*, which revolves daily from east to west, and carries along with it all the other orbs. Above the whole spreads the *Grand Empyrean*, or third heavens, the abode of perpetual serenity.*

To account for the planetary motions, it was supposed that each of the planetary orbs, as well as that of the sun, has a motion of its own eastward, while it partakes of the common diurnal motion of the starry sphere. Aristotle taught that these motions are effected by a tutelary genius of each planet, residing in it, and directing its motions, as the mind of man directs his motions.

442. On coming down to the time of Hipparchus, who flourished about 150 years before the Christian era, we meet with astronomers who acquired far more accurate knowledge of the celestial motions. Hipparchus was in possession of instruments for measuring angles, and knew how to resolve spherical triangles. He ascertained the length of the year within 6m. of the truth. He discovered the eccentricity of the solar orbit (although he supposed the sun actually to move uniformly in a circle, but the earth to be placed out of the center), and the positions of the sun's apogee and perigee. He formed very accurate estimates of the obliquity of the ecliptic and of the precession of the equinoxes. He computed the exact period of the synodic revolution of the moon, and the inclination of the lunar orbit; discovered the motion of her node and of her line of apsides; and made the first attempts to ascertain the horizontal parallaxes of the sun and moon.

Such was the state of astronomical knowledge when Ptolemy wrote the Almagest, in which he has transmitted to us an encyclopædia of the astronomy of the ancients.

* Long's Ast., ii., p. 640; Robinson's Mech. Phil., ii., p. 83; Gregory's Ast. p. 132; Playfair's Dissertations, p. 118.

443. The systems of the world which have been most celebrated are three—the Ptolemaic, the Tychonic, and the Copernican. We shall conclude this part of our work with a concise statement and discussion of each of these systems of the Mechanism of the Heavens.

THE PTOLEMAIC SYSTEM.

444. The doctrines of the Ptolemaic System were not originated by Ptolemy; but being digested by him out of materials furnished by various hands, it has come down to us under the sanction of his name.

According to this system, the earth is the center of the universe, and all the heavenly bodies daily revolve around it from east to west. In order to explain the planetary motions, Ptolemy had recourse to *deferents* and *epicycles*,—an explanation devised by Apollonius, one of the greatest geometers of antiquity.* He conceived that, in the circumference of a circle, having the earth for its center, there moves the center of another circle, in the circumference of which the planet actually revolves. The circle surrounding the earth was called the *deferent*, while the smaller circle, whose center was always in the periphery of the deferent, was called the *epicycle*. The motion in each was supposed to be uniform. Lastly, it was conceived that the motion of the center of the epicycle in the circumference of the deferent, and of the deferent itself, are in opposite directions, the first being toward the east, and the second toward the west.

445. But these views will be better understood from a diagram. Therefore, let ABC (Fig. 81) represent the *deferent*, E being the earth a little out of the center. Let *abc* represent the *epicycle*, having its center at *v*, on the periphery of the deferent. Conceive the circumference of the deferent to be carried about the earth every twenty-four hours in the order of the letters; and at the same time, let the center *v* of the epicycle *abcd*, have a slow motion in the opposite direction, and let a body revolve in this circle in the direction *abcd*. Then it will

* Playfair, Dissertation Second, p. 119.

be seen that the body would actually describe the looped curves *klmnop*; that it would appear stationary at *l* and *m*, and at *n* and *o*; that its motion would be direct from *k* to *l*, and then retrograde from *l* to *m*; direct again from *m* to *n*,

Fig. 81.

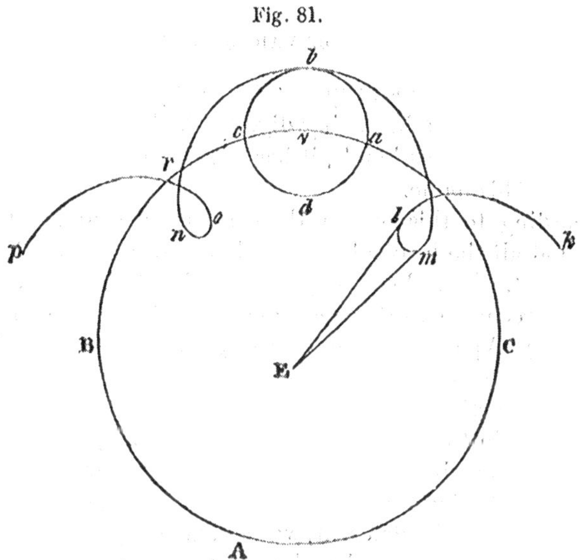

and retrograde from *n* to *o*. Thus, suppose Mercury to be situated at *b* in its epicycle. By the revolution of the deferent, it would be carried along with the other heavenly bodies around the earth from left to right, every twenty-four hours; but, meanwhile, the center of the epicycle shifting its place slowly from right to left, while Mercury was moving from *b* to *c*, *c* itself would change its place to *r*, and therefore the path of the planet would be in the cycloidal arc *br*. Again, while Mercury was passing through *cda*, the point *c* would be still moving eastward, which would have the effect apparently to compress the lower half of the epicycle into the looped curve *nor*; and as on this side the motion in the epicycle is in the same direction with that of the deferent, but at a slower rate, the apparent path is much shorter than where, as on the other side, the two motions conspire.

446. Such a deferent and epicycle may be devised for each planet as will fully explain all its ordinary motions; but it is inconsistent with the *phases* of Mercury and Venus, which being between us and the sun on both sides of the epicycle, would present their dark sides toward us in both these positions, whereas at one of the conjunctions they are seen to shine with full face.* It is moreover absurd to speak of a geometrical center, which has no bodily existence, moving around the earth on the circumference of another circle; and hence some suppose that the ancients merely assumed this hypothesis as affording a convenient geometrical representation of the phenomena—a diagram simply, without conceiving the system to have any real existence in nature.

447. The *objections* to the Ptolemaic system, in general, are the following: First, it is a mere hypothesis, having no evidence in its favor, except that it explains the phenomena. This evidence is insufficient of itself, since it frequently happens that each of two hypotheses, directly opposite to each other, will explain all the known phenomena. But the Ptolemaic system does not even do this, as it is inconsistent with the phases of Mercury and Venus, as already observed. Secondly, now that we are acquainted with the distances of the remoter planets, and especially of the fixed stars, the *swiftness of motion* implied in a daily revolution of the starry firmament around the earth, renders such a motion wholly incredible. Thirdly, the *centrifugal force* that would be generated in these bodies, especially in the sun, renders it impossible that they can continue to revolve around the earth as a center.

These reasons are sufficient to show the absurdities of the Ptolemaic System of the World.

THE TYCHONIC SYSTEM.

448. Tycho Brahe, like Ptolemy, placed the earth in the center of the universe, and accounted for the diurnal motions in the same manner as Ptolemy had done, namely, by an actual revolution of the whole host of heaven around the earth every twenty-four hours. But he rejected the scheme of deferents

* Vince's Complete System, i., p. 96.

and epicycles, and held that the moon revolves about the earth as the center of her motions; that the sun, and not the earth, is the center of the planetary motions; and that the sun, accompanied by the planets, moves around the earth once a year, somewhat in the manner that we now conceive of Jupiter and his satellites as revolving around the sun. The system of Tycho serves to explain all the common phenomena of the planetary motions, but it is encumbered with the same objections as those that have been mentioned as resting against the Ptolemaic system, namely, that it is a mere hypothesis; that it implies an incredible swiftness in the diurnal motions; and that it is inconsistent with the known laws of universal gravitation. But if the heavens do not revolve, the earth must, and this brings us to the system of Copernicus.

THE COPERNICAN SYSTEM.

449. Copernicus was born at Thorn, in Prussia, in 1473. The system that bears his name was the fruit of forty years of intense study and meditation upon the celestial motions. As already mentioned (Art. 6), it maintains (1), That the *apparent* diurnal motion of the heavenly bodies, from east to west, is owing to the *real* revolution of the earth on its own axis from west to east; and (2), That the sun is the center around which the earth and planets all revolve from west to east. It rests on the following arguments:

First, *the earth revolves on its own axis.*

1. Because this supposition is vastly more *simple.*

2. It is agreeable to *analogy*, since all the other planets that afford any means of determining the question, are seen to revolve on their axes.

3. The *spheroidal figure* of the earth is the figure of equilibrium, that results from a revolution on its axis.

4. The *diminished weight* of bodies at the equator, indicates a centrifugal force arising from such a revolution.

5. Bodies let fall from a high eminence, fall *eastward of their base*, indicating that higher objects have greater velocity of rotation than lower ones.

6. The *precession of the equinoxes* is explained by the earth's rotation on its axis.

Secondly, *the planets, including the earth, revolve about the sun.*

1. The *phases* of Mercury and Venus are precisely such as would result from their circulating around the sun in orbits within that of the earth; but they are never seen in opposition, as they would be if they circulated around the earth.

2. The superior planets do indeed revolve around the earth; but they also revolve around the sun, as is evident from their phases and from the known dimensions of their orbits; and that the sun, and not the earth, is the *center* of their motions, is inferred from the greater symmetry of their motions as referred to the sun than as referred to the earth, and especially from the laws of gravitation, which forbid our supposing that bodies so much larger than the earth, as some of these bodies are, can circulate permanently around the earth, the latter remaining all the while at rest.

3. The annual motion of *the earth* itself is indicated also by the most conclusive arguments. For, first, since all the planets with their satellites, and the comets, revolve about the sun, *analogy* leads us to infer the same respecting the earth and its satellite. Secondly, the motions of the satellites, as those of Jupiter and Saturn, indicate that it is a law of the solar system that the smaller bodies revolve about the larger. Thirdly, the direction of the periodical meteors of November, which, in a majority of cases, is from east to west, indicates the motion of the earth from west to east. Lastly, the aberration of light affords a sensible proof of the motion of the earth, since that phenomenon indicates both a progressive motion of light, and a motion of the earth from west to east. (Art. 195.)

450 It only remains to inquire whether there subsist higher orders of relations between the stars themselves. The assemblage of bodies in *clusters*, as in the Pleiades, and still more, as in the great nebula of Hercules (Art. 425), implies mutual relations constituting for each a system within itself; and the analogies of all that portion of the heavenly bodies, whose motions fall within our observation, and the known uniformity of the laws of nature, conspire to prove that those relations are maintained by revolutions around a common center. What theory would lead us to expect, we actually see exemplified in the revolutions of the *binary stars* (Art. 430), and in the motion of the sun himself with his attendant worlds (Art. 432). The *Nebulæ* also compose peculiar systems, in which the members seem associated in mutual relations, and separated from all the other heavenly bodies, each composing an "island universe." Thus we ascend from the lower to the higher combinations, according to a *uniform plan*, so characteristic of ascending orders in every department of nature. Beginning with the relation between the earth and its satellite, we see it sustained by the prevalence of forces which subject it to Kepler's laws and the law of universal gravitation. We see the same principles carried out on a larger scale, but exactly on the same *plan*, in the system of Jupiter and his satellites, and in the respective systems of Saturn, Uranus, and Neptune. From this lowest order of combination, composed of planets and their satellites, we ascend to the next higher order, consisting of suns and planets, in which the same plan is exemplified on a still grander scale, but without any change in its peculiar features. We next ascend still higher to the third order, as in the *binary stars*, where sun revolves around sun, upon the same unvarying plan as before seen in these nearer worlds. At present, *observation* leads us to no higher point of the scale in the structure of the universe; but the mind of man, obtaining from these lower systems a knowledge of the plan on which the universe is built, goes forward to complete the grand machine. A bold attempt has recently been made by *Maedler*, an eminent European astronomer, to fix the center, around which not only our sun, but all the stars of our firmament revolve. It must evidently be such a point, that the known proper motions detected among the fixed stars will conform to it, like the motions of the planets around the sun. He places that center in the Pleiades, or, more exactly, in *Alcyone*, the central star of the Pleiades, which body is therefore denominated the CENTRAL SUN.* The proofs of this remarkable hypothesis are deemed too incomplete at present to command entire assent; but the method of investigation pursued by this distinguished astronomer, opens a new field of observation and of speculation, and promises to lend a new interest to inquiries into the mechanism of the universe.

---

* Plate III., 1.

451. This fact being now established, that the stars are immense bodies like the sun, and that they are subject to the laws of gravitation, we can not conceive how they can be preserved from falling into final disorder and ruin, unless they move in harmonious concert like the members of the solar system. Otherwise, those that are situated on the confines of creation, being retained by no forces from without, while they are subject to the attraction of all the bodies within, must leave their stations, and move inward with accelerated velocity, and thus all the bodies in the universe would at length fall together in the common center of gravity. The immense distances at which the stars are placed from each other would indeed delay such a catastrophe; but such must be the ultimate tendency of the material world, unless sustained in one harmonious system by nicely-adjusted motions.* To leave entirely out of view our confidence in the wisdom and preserving goodness of the Creator, and reasoning merely from what we know of the stability of the solar system, we should be justified in inferring that other worlds are not subject to forces which operate only to hasten their decay, and to involve them in final ruin.

We conclude, therefore, that the material universe is one great system; that the combination of planets with their satellites constitutes the first or lowest order of worlds; that next to these, planets are linked to suns; that these are bound to other suns, composing a still higher order in the scale of being; and, finally, that all the different systems of worlds move around their common center of gravity.

---

* Robison's Physical Astronomy.

PLATE I.

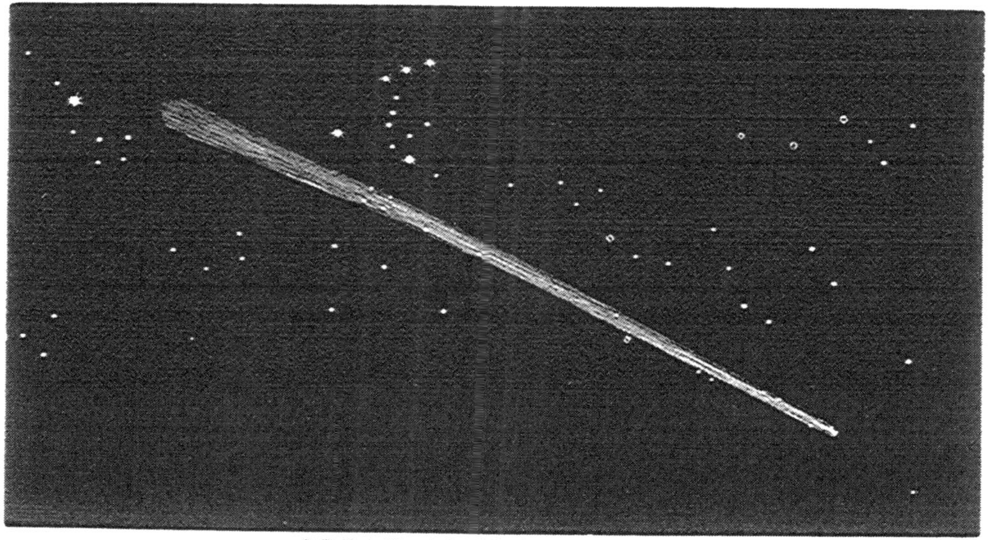

GREAT COMET OF 1843.
AS SEEN AT YALE COLLEGE, MARCH 20TH.

PLATE II.

NEBULA AND DOUBLE STARS.

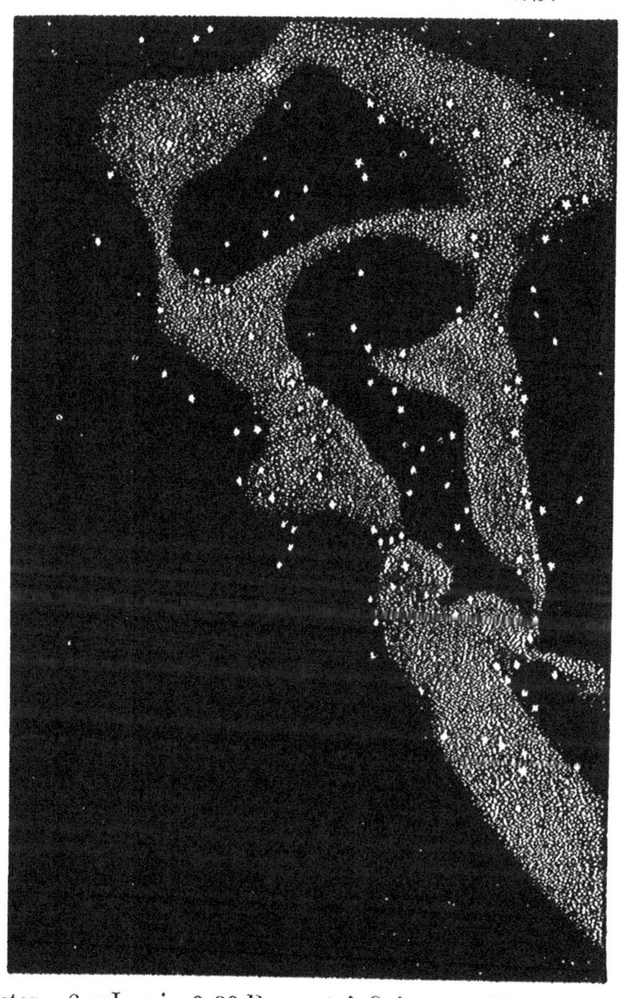

1. Castor. 2. γ Leonis. 3. 39 Drac. 4. λ Oph. 5. 11 Monoc. 6. ζ Cancri.

s

Revolutions of γ Virginis.

n

1837. 1838. 1839. 1840. 1845. 1850. 1860. Orbit.

PLATE III.
CLUSTERS AND NEBULÆ.

www.ingramcontent.com/pod-product-compliance
Lightning Source LLC
LaVergne TN
LVHW010206110826
845151LV00002B/618

* 9 7 8 1 4 2 5 5 3 5 1 6 2 *